SALTSCAPES

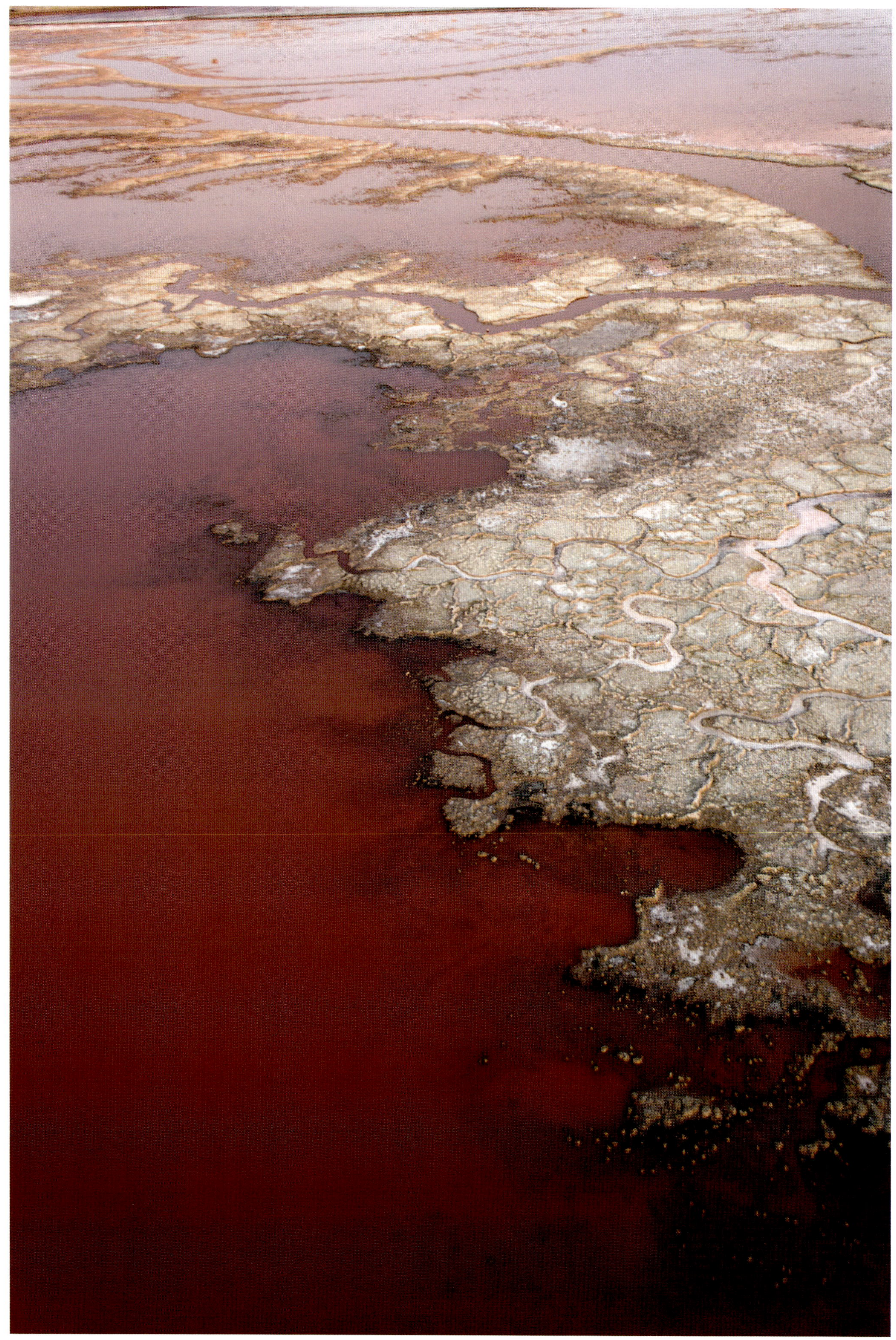

Plate 1. Red pond at sunset (Salt Pond M5, near Mowry Slough, July 2008)

SALTSCAPES

THE KITE AERIAL PHOTOGRAPHY OF CRIS BENTON

Cris Benton

Foreword by Will Travis

HEYDAY, BERKELEY, CALIFORNIA

To my grandparents, Otis and Sarah Knight, who by fine example instilled a
lifelong appreciation of curiosity and experimentation

* * *

The author and publisher are grateful to Cargill for its generosity and gracious cooperation at every stage of this project.

Library of Congress Cataloging-in-Publication Data

Benton, Cris.
  The kite aerial photography of Cris Benton / Cris Benton ; foreword by Will Travis.
     pages cm
  Includes bibliographical references.
  ISBN 978-1-59714-247-2 (hardcover : alk. paper)
  1. San Francisco Bay Area (Calif.)--Aerial photographs. 2. Aerial photography--California--San Francisco Bay Area. 3. Kites. I. Title.
  F869.S343B46 2013
  917.94′60454--dc23
                        2013010163

Cover photograph: Sanguine shallows (Salt Pond M5, July 2008)
Book design by Lorraine Rath
Map on page v by Kyle Dahl at Marble Environmental Imaging. 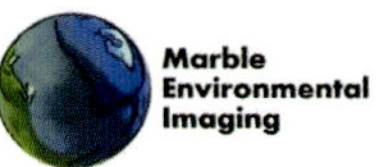

Map Data: US Geological Survey, National Atlas of the United States (active salt ponds, Cargill salt plant site, surface water), San Francisco Estuary Institute (former salt ponds, restoration area boundaries), California Department of Transportation (roads), Oak Ridge National Laboratory (railroads)

Orders, inquiries, and correspondence should be addressed to:
    Heyday
    P.O. Box 9145, Berkeley, CA 94709
    (510) 549-3564, Fax (510) 549-1889
    www.heydaybooks.com

Printed in China by Everbest Printing Co. through Four Colour Imports, Ltd., Louisville, Kentucky

10   9   8   7   6   5   4   3   2   1

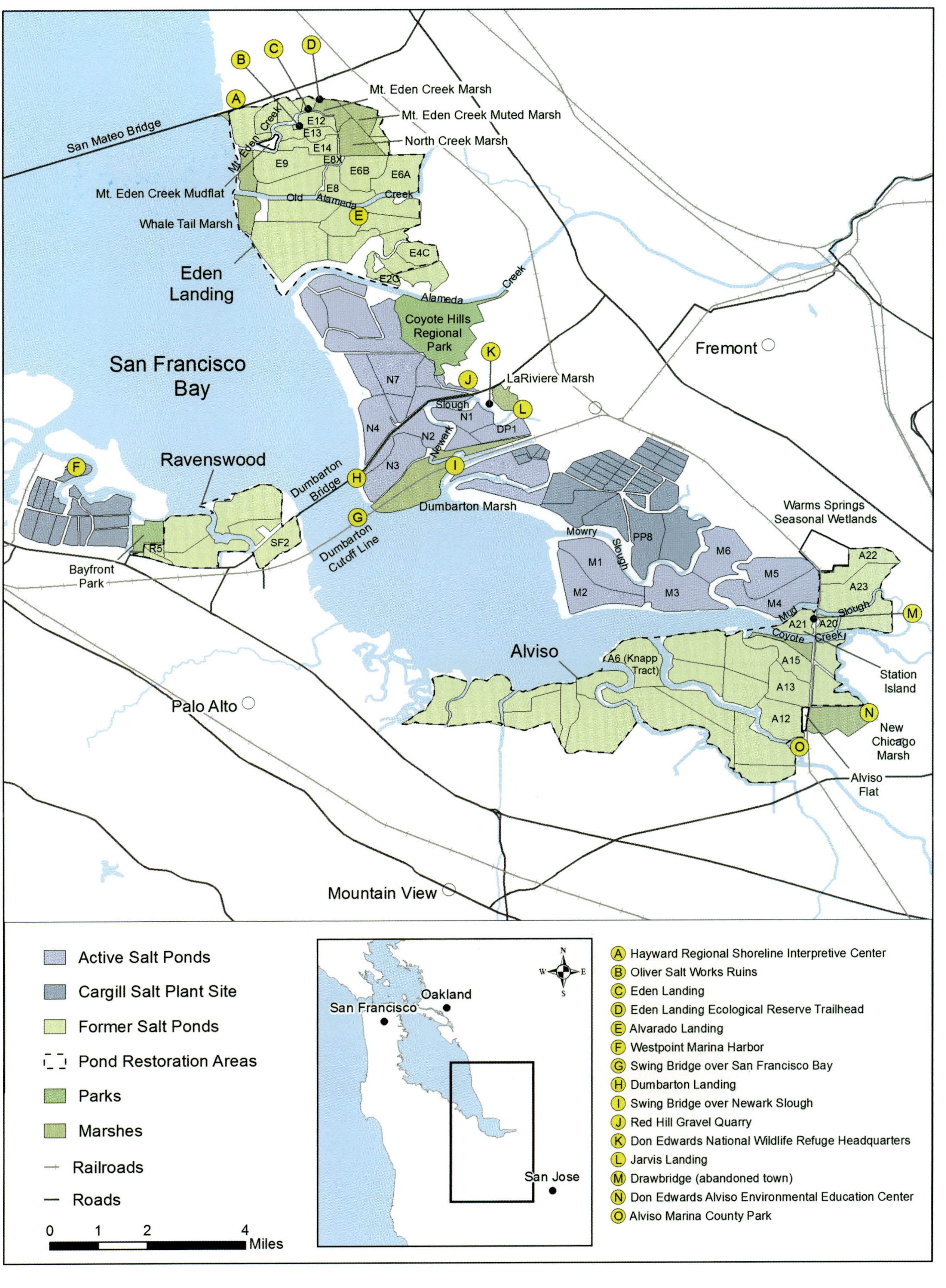

San Mateo Bridge
Mt. Eden Creek Marsh
Mt. Eden Creek Muted Marsh
North Creek Marsh
E12
E13
E14
E9
E8X
E6B
E6A
Mt Eden Creek
Old Alameda Creek
E8
Mt. Eden Creek Mudflat
Whale Tail Marsh
Eden Landing
E4C
E2C
San Francisco Bay
Alameda Creek
Coyote Hills Regional Park
Fremont
N7
LaRiviere Marsh
Slough
N1
DP1
N4
N2
Newark
Ravenswood
N3
Dumbarton Bridge
Warms Springs Seasonal Wetlands
F
Dumbarton Marsh
A22
Mowry
PP8
M6
A23
R5
SF2
Dumbarton Cutoff Line
M1
Slough
M5
M
Bayfront Park
M2
M3
M4
Mud
Coyote Creek
Slough
A21
A20
Alviso
A6 (Knapp Tract)
A15
Station Island
Palo Alto
A13
N
A12
New Chicago Marsh
O
Alviso Flat
Mountain View

Active Salt Ponds
Cargill Salt Plant Site
Former Salt Ponds
Pond Restoration Areas
Parks
Marshes
Railroads
Roads

0   1   2   4 Miles

N
W   E
S

Oakland
San Francisco
San Jose

A Hayward Regional Shoreline Interpretive Center
B Oliver Salt Works Ruins
C Eden Landing
D Eden Landing Ecological Reserve Trailhead
E Alvarado Landing
F Westpoint Marina Harbor
G Swing Bridge over San Francisco Bay
H Dumbarton Landing
I Swing Bridge over Newark Slough
J Red Hill Gravel Quarry
K Don Edwards National Wildlife Refuge Headquarters
L Jarvis Landing
M Drawbridge (abandoned town)
N Don Edwards Alviso Environmental Education Center
O Alviso Marina County Park

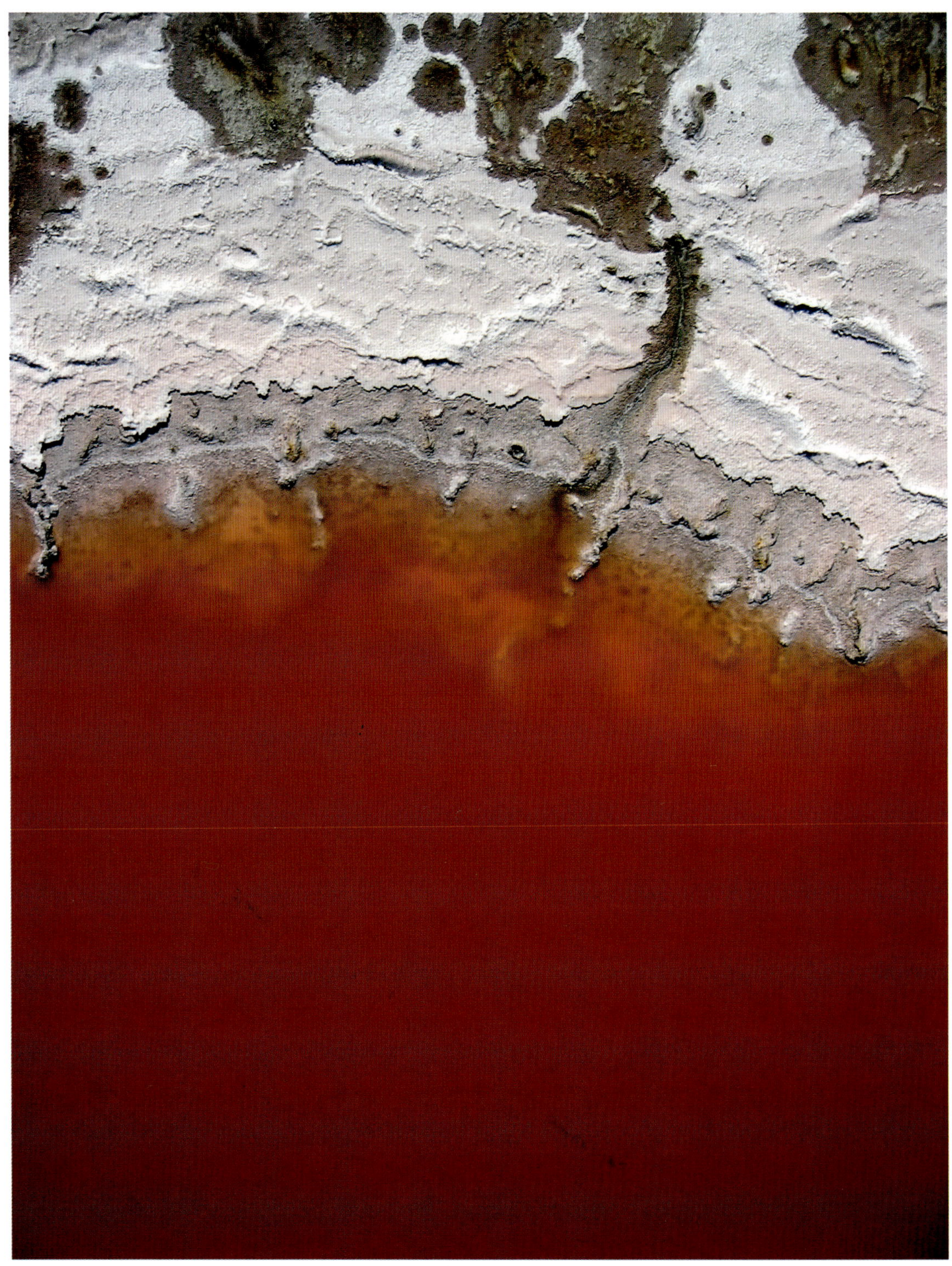

Plate 2. Salt-pond shoreline (Salt Pond N1, August 2003)

CONTENTS

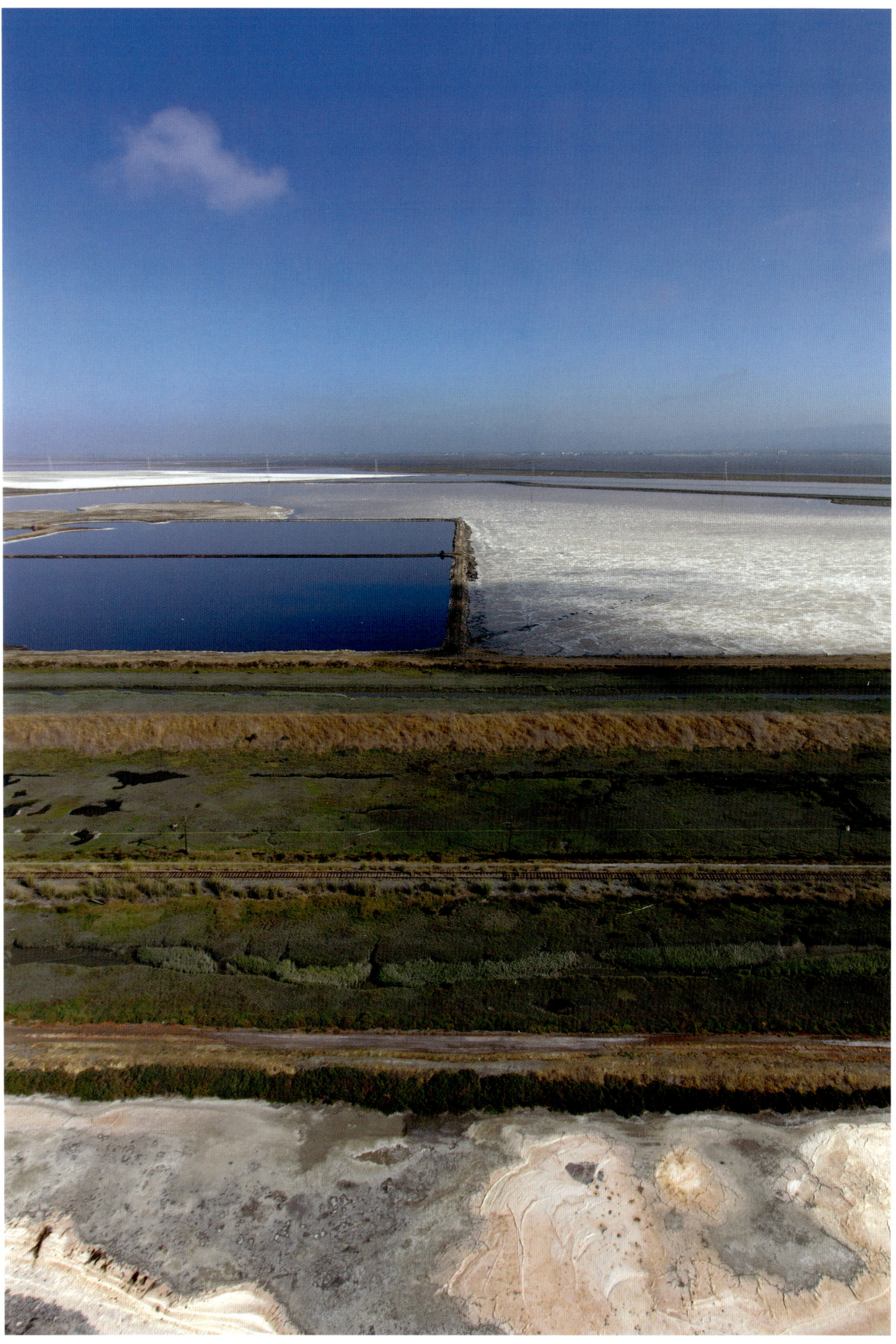

Plate 3. Strip of constrained marsh (Dumbarton Marsh, July 2012)

FOREWORD

Will Travis

For three decades I worked for the San Francisco Bay Conservation and Development Commission, the California state agency created in 1965 to "save the bay" from the ongoing land reclamation projects that had reduced the size of the bay by a third over the previous century. My day-to-day activities ranged from the mundane chores inherent in the bureaucratic operation of any government agency to the headline-grabbing political victories that came from dealing successfully with challenging policy issues like climate change. But my most enduring pleasure and satisfaction came from two things: having the opportunity to experience the many facets of the bay's physical beauty over time; and getting to know so many of the wide range of people whose lives are intrinsically linked to the bay.

Some might be surprised that I value the aesthetics of the bay more than its contribution as an environmental treasure or an economic asset. At first placing such a high value on beauty may sound trivial. After all, beauty is ephemeral. It would seem to have no value. But, in fact, the bay's beauty does have value. A house with a view of the bay costs more. Visitors pay a premium price for a hotel room facing the bay. People will spend more for a mediocre dinner in a restaurant overlooking the bay than for a sensational meal in a restaurant that's not on the waterfront.

Sylvia McLaughlin, one of the pioneers in the "save the bay" movement, shares my view. She once told me, "We didn't save the bay to improve water quality or protect wetlands. We saved it because we had never seen anything so beautiful." In short, the bay was saved to keep it pretty.

The bay's beauty is the underpinning of its spiritual value. The bay is the heart, soul, and essence of our region. The bay gives our region its name and creates the sense of place that defines the community where we live. This community encompasses a wide assortment of characters including the architects, engineers, lawyers, developers, and financiers who crafted the deals to build houses, hotels, offices, bridges, highways, and a wide array of other projects, as well as the government regulators and environmentalists who zealously pored over the details of those deals so the bay wouldn't be harmed by the new development. Interacting with these people enriched my life and made me deeply appreciate that each of them, in their own way, is a steward of San Francisco Bay.

Cris Benton's photography of the bay embodies both my love of the bay's beauty and my fascination with the people who share my interconnectedness with the bay. Cris is a member of the zany cast of characters who make up the family of bay stewards. His credentials include being an emeritus professor of architecture at the University of California, Berkeley. But at heart, he's a playful kid, a tinkerer who has combined his hobby of flying remote-controlled model airplanes with his talent as a photographer by figuring out how to mount a remote-controlled camera on a kite. His kites fly higher than any ground-based boom can reach and lower than any aircraft can quietly fly, an altitude that gives Cris's work its own unique range of perspectives. Cris doesn't rely on a video monitor to see what his kite-borne camera is seeing. He looks up at the kite, looks at which way the camera is pointing, and in his mind's eye sees how the photo will turn out. He's sometimes surprised at the details, but seldom does he have to crop or edit the image. In my mind, this capability differentiates Cris from a photojournalist; he's a superb artist.

I am most enchanted with those of Cris Benton's photos that could well be pictures of Mark Rothko paintings. Only on close examination do we discover that we are looking at an aerial photo of tractor tire tracks across a muddy stretch of bay bottom. Others have photographed the Mondrian-like mosaic of multi-colored salt ponds, but Cris brings life to the patterns and allows us to see what is going on beneath the shapes and tones. Through Cris's work I've found a portal into an aesthetic perspective of the bay that I never before experienced.

Cris's photos chronicle one of the greatest landscape-scale transformations underway in America: tens of thousands of acres of ponds used in the process of making salt are being restored to natural habitat. Salt, which evaporated naturally from shallow bay waters, was long harvested by Native Americans who lived along the bay shoreline. Later, Spanish and Mexican inhabitants gathered salt out of natural tide pools. The salt-making industry started in the 1850s when a number of companies began producing salt for the emerging gold and silver mining operations in California and Nevada. The salt companies built levees and dikes to create shallow ponds along the north and south ends of the bay, often destroying historic marshes and wetlands. In the early twentieth century, Leslie Salt Company began its operations. Over the next half-century Leslie purchased a number of smaller salt companies, and by mid-century the company owned about sixty-four square miles of property under the bay and along its shoreline. In 1978 Cargill, a multinational corporation, purchased Leslie.

In the 1960s, citizens concerned about the continuing loss of bay wetlands began working to protect the existing marshlands and seek the restoration of salt ponds to natural habitat. They found an ally in Congressman Don Edwards, and through their tireless work, federal legislation was passed establishing the Don Edwards San Francisco Bay National Wildlife Refuge as the first and largest urban wildlife refuge in the United States, near San Jose. Over the following decades the boundaries of the refuge have been expanded, and many other organizations and government agencies have joined the salt-pond restoration campaign. Cargill sold ten thousand acres of North Bay salt ponds to public agencies in 1994. This public acquisition, along with another in 2003 when the company sold fifteen thousand acres of South Bay salt ponds to the public, was possible only because Cargill sold the land well below its fair market value; without this donation, it's unlikely the public agencies could have afforded to buy the property.

The work that's featured in this book focuses on the southern end of the bay, where the conversion of the former salt ponds is now underway. Some of the ponds are being restored to marshes and wetlands. Others will remain the types of shallow bay ponds that have become habitat for a variety of wildlife species over the past century. At the same time, flood protection for Silicon Valley, which borders the South Bay salt ponds, is being incorporated into the long-range plans for this landscape transformation. Public wildlife managers, engineers, and planners are in charge, but Cargill's employees remain active participants in the process they unleashed through their caring stewardship of their property. The company's employees are intimately familiar with the hydrology of the bay and the vagaries of nature that only someone who works hands-on in the bay's wetlands can know. Cargill employees applied this knowledge in the North Bay by volunteering their time to assist the staff at the California Department of Fish and Wildlife who had taken on the task of converting the salt ponds to wetlands. In the South Bay, the *Mallard II*, the dredge used by Cargill to maintain the levees between the salt ponds, is an officially designated historic structure which its staff operates according to tight restrictions.

In addition to the sheer scale of this transformation, it is remarkable for the unprecedented pace at which it is being carried out. Around the nation, we can find other examples of wonderful reuse of once vibrant industrial facilities. The High Line in New York City and the Sands Casino in Bethlehem, Pennsylvania, are examples. A linear park now stretches above Manhattan neighborhoods where an elevated railroad once operated. And a gaming resort occupies the massive structures that were once filled with blast furnaces and molten steel. But in both of these examples and most others, the industrial facilities were abandoned and suffered decades of decay before they were found worthy of reuse. In the Bay Area, the transformation is taking place instantly. Salt-making operations continue on land that is part of a wildlife refuge, while other salt ponds are being rededicated to natural habitat. This unique conversion process is being made possible by the close partnership between Cargill, many public agencies, and caring citizens.

Cris Benton's photos graphically illustrate this massive transformation process with great sensitivity. He is as respectful of the stark beauty of a rail line and the elegance of a pile of white salt as he is of the softness of a snowy plover and the emergence of marsh grass from a mudflat. Enjoy this book as a collection of works of art. Or enjoy it as the story of an historic transformation. You can do both because of Cris Benton's genius and the dedication of the people who have brought this transformation to reality.

Will Travis served as the executive director of the San Francisco Bay Conservation and Development Commission from 1995 to 2011.

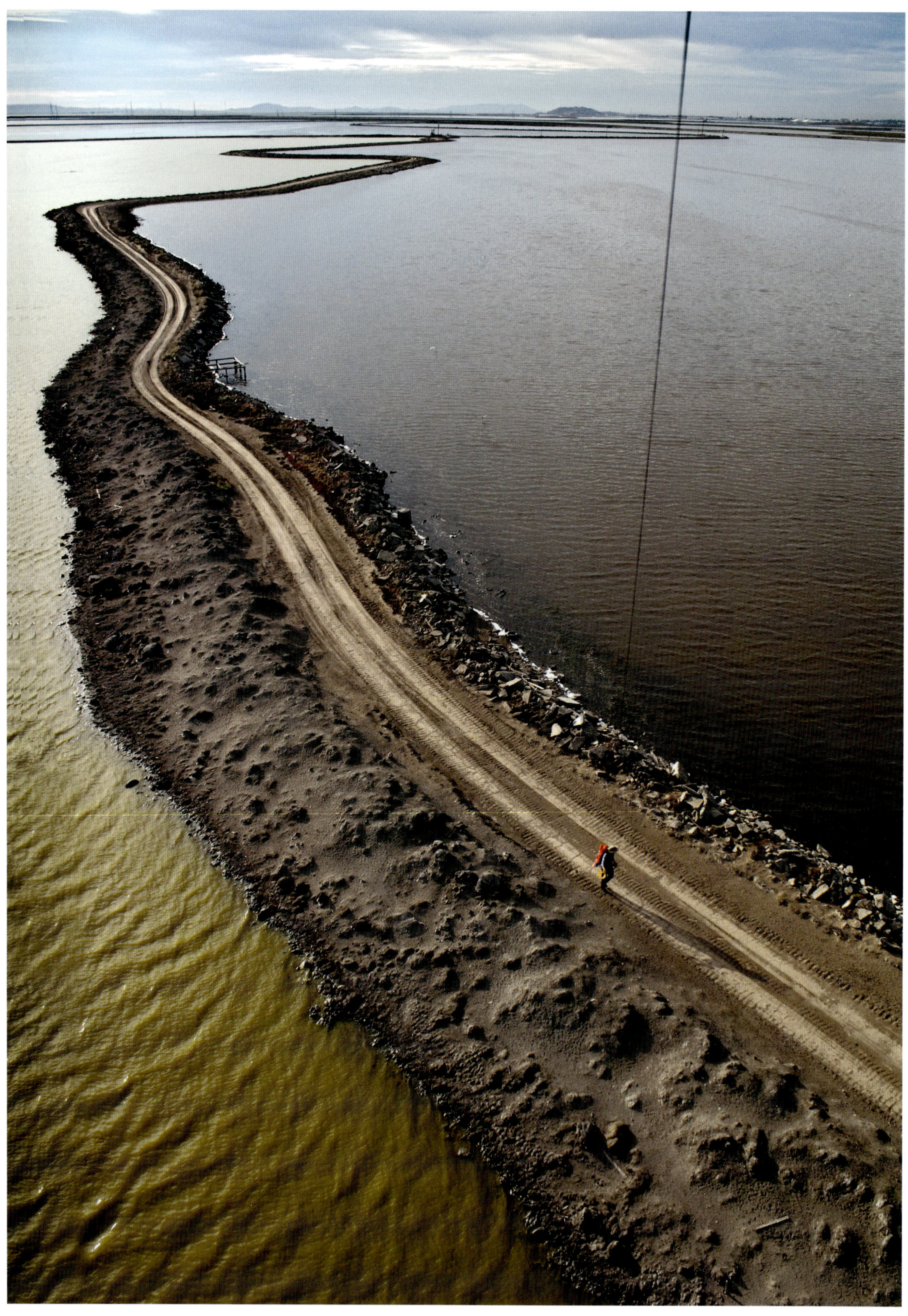

Plate 4. Levee marking course of former slough (Salt Ponds A13 and A15, August 2009)

1: KITE PLAY

 to the exuberant, otherworldly landscape of South San Francisco Bay. There, depending on the mood of a Sunday, I can bring binoculars to bear on the strikingly abundant wildlife, explore diverse halophilic microorganisms with a field microscope, hike out to ponder early engineering interventions scattered across the bay shallows, or (my favorite) launch a kite-lofted camera to photograph juxtapositions in the landscape from above. And juxtapositions abound—dendritic marsh channels as foils for the straight lines of infrastructure; wild openness confronting the creeping front of capitalism; salt ponds, vividly colored by the aforementioned halophiles, constrained by subtly hued mud and marsh; derelict, forgotten saltworks faintly echoing their former functions.

Over time my curiosity about the South Bay landscape became a sustained fascination. For behind the visual richness of these juxtapositions lie the South Bay's interesting history and the active formulation, at this very moment, of bold initiatives for its future. So, for ten years now I have hiked the salt-pond levees and taken low-level aerial photographs using cameras lofted by kites. That these images are often visually striking is in no small part because they reveal hidden and often enigmatic aspects of the landscape. From relatively low altitudes of three feet to three hundred feet above the ground, photographs taken with a kite offer a view from just beyond normal human experience, sometimes by only a few feet. These low altitudes, prime and largely unexplored territory in aerial photography, yield images intimate in detail and form. It is a vantage point ideally suited to the South Bay, for it greatly reduces sky reflection from the salt-pond surfaces to expose colors, textures, patterns, and traces invisible from the ground.

The kite aerial photographs provided a particularly engaging way to see South San Francisco Bay in what became an iterative exploration. The process began with the South Bay as a terra incognita of sorts—a vast, vague wetland to be skirted or crossed while heading somewhere else. My early images along the periphery led to obvious questions about the landscape's colors and textures, while later sessions piqued curiosity about more subtle features. In what became an enjoyable and repeated pattern, questions raised by the photographs would spawn visits to the library, conversations with South Bay folks, map browsing, countless hours with Google Earth, and bouts of speculation. Then it was back to the field for new photographs, where for each question answered new questions emerged. In the process, still ongoing, I have developed a great affection for this remarkable landscape and an appreciation of its transitional nature.

tend to stumble into things and the South Bay is no exception. Years ago I flew radio-controlled sailplanes, an entertaining way to engage Bay Area topography and meteorology. One day while flying a glider at Berkeley's Cesar Chavez Park, I had a chance conversation with Anne Rock, a Berkeley resident who used similar radio-control gear not for airplanes but to control kite-lofted cameras. Having been an avid photographer for decades, I immediately resonated with this idea. My next winter break was devoted to designing and building a rudimentary apparatus to loft a 35mm compact camera so that I might take my own kite aerial photographs.

Inclement weather and academic duties delayed the maiden flight of my kite camera until a fine spring day in March 1995. With great anticipation I loaded film, launched the kite, and took a round of photographs along the section of Berkeley waterfront where I had flown sailplanes on dozens of occasions. On viewing the photographs a few days later I was irreversibly seduced. Here were views of the landscape seen vicariously as though from my sailplane. More importantly, they revealed aspects of the place that I had not previously noticed such as a wonderful pattern of animal tracks through the underbrush.

Kite aerial photography (KAP) appeals to that part of me, perhaps of all of us, that would slip earthly bonds and see the world from new heights. Its aerial views offer a fresh perspective of familiar landscapes and in doing so challenge my spatial sensibilities, my grasp of relationships. The technique fuels my sustained fascination with photography's capacity to reveal patterns and phenomena that lie beyond the capacities of our native senses—think Muybridge's motion studies, Edgerton's frozen moments, and the Eameses' decadal plunge through scale.

Thus began my career in kite aerial photography. Almost two decades later I think of my KAP endeavors as having three phases, each roughly equal in duration. The first phase focused on the mechanics of KAP—learning about kites, refining camera cradles, studying urban aerodynamics, and developing strategies for camera survival. These were happy days for a tinkerer such as myself. The middle phase revolved around a quest for compositionally satisfying images. For years, my KAP gear traveled with me in a small backpack, and whenever possible I would work a setting for aerial images. This phase built valuable experience flying kites and produced a playful, eclectic collection of photographs. In the third and current phase, KAP has become a means to explore and document specific landscapes. Here technique yields primacy to subject and KAP serves as a means of inquiry in a sustained examination of place. Each phase has built on its predecessor, with the foundation being the mechanics of KAP.

The keyword for my first phase of kite photography was invention. Taking photographs from kites is a straightforward enough idea. Select a reliable kite with sufficient pull and fly it to an altitude where airflow is smooth. After testing the kite's flight for a few minutes to confirm stability, attach a camera, housed in a suspended cradle, to the kite line. The camera is at least a hundred feet from the kite, a separation that contributes to stable flight. Play out more kite line and the kite flies higher, pulling the camera cradle after it. When the camera has reached a desired altitude, compose the shot by aiming the camera with a remote control and fire the shutter. To work this ground I am using techniques borrowed from a century ago, well before airplanes carried cameras, when earthbound photographers used great ingenuity to send cameras skyward.

From the very beginning of photography inventors were eager to get cameras airborne. In the decades before fixed-wing aircraft became practical, photographers used balloons, rockets, pigeons, and kites to capture aerial views. The first kite aerial photographs are attributed to Arthur Batut of Labruguière, France, who employed a home-built rig in 1887. Batut's camera was mounted to the kite itself along with an altimeter situated to encode the exposure altitude on the

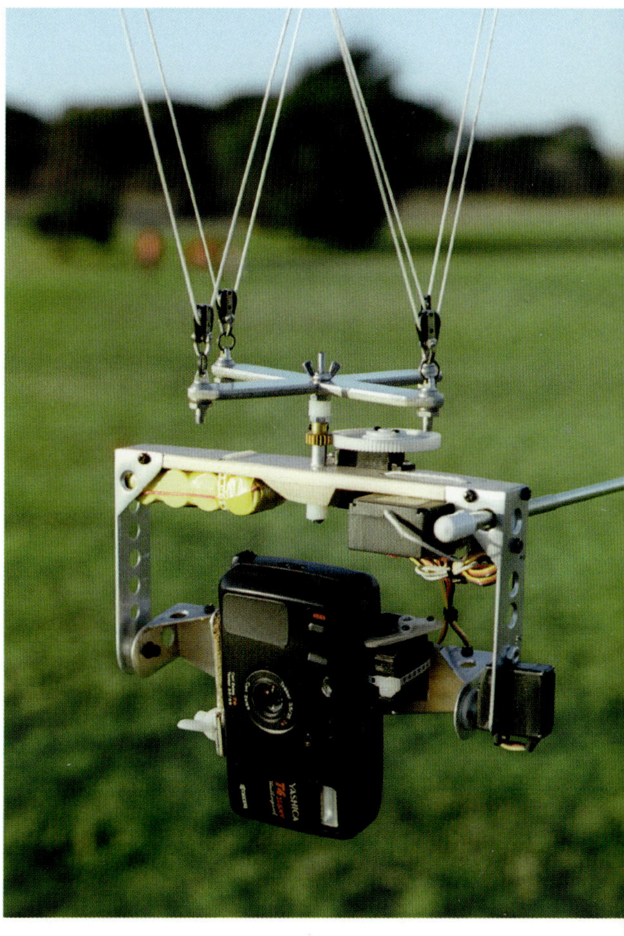

One of my early camera cradles carrying the very capable Yashica T4 35mm point-and-shoot camera. The cross at the top with small blocks is a Picavet suspension.

film. A slow-burning fuse actuated the shutter a fixed amount of time after the kite was launched, typically a few minutes. The shutter release dropped a small flag to indicate it was time to haul the kite down. In 1906 George Lawrence produced the most famous of the early kite aerial photographs when he captured an aerial view of San Francisco after the earthquake using a fifty-pound, swing lens panoramic camera. This period of energetic invention yielded devices like Pierre Picavet's 1912 string-and-pulley camera suspension, a technique I use today.

By World War I fixed-wing aircraft had become the normative platform for aerial photography with kites, balloons, and pigeons, appearing in isolated and idiosyncratic citations—archaeology in the Sudan, a *Popular Mechanics* article in the Depression. Then the emergence of the Internet in the mid-1990s allowed kite aerial photography to blossom again as interested parties, largely hobbyists, found each other to share technique and photographic adventure. Our day's modest resurgence of KAP was further fueled by modern materials for kite building, accessible technologies for radio control, and the automated features of modern cameras.

Contemporary kite aerial photographers, or KAPpers as we call ourselves, use two basic techniques. The first, AutoKAP, employs timers and electronic circuits to rotate and fire the camera automatically at timed intervals. The second involves radio-controlled cradles remotely manipulated

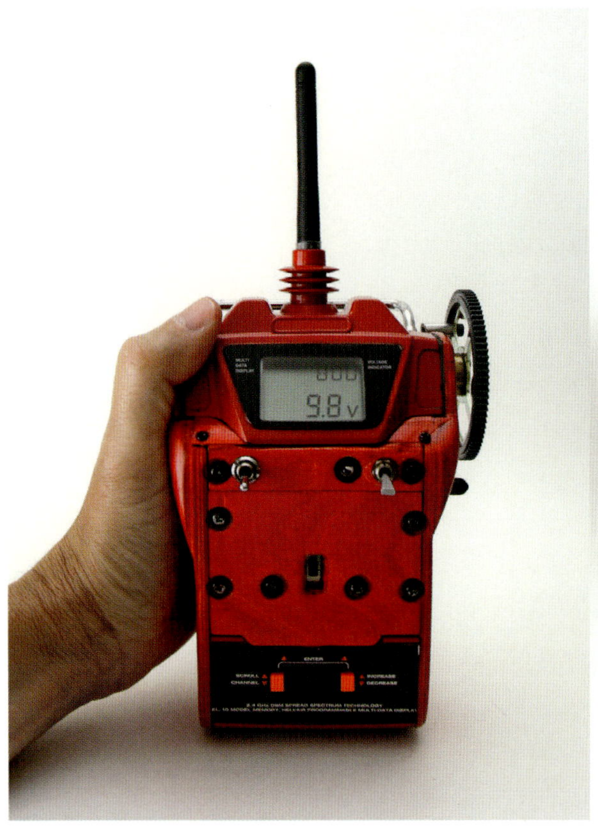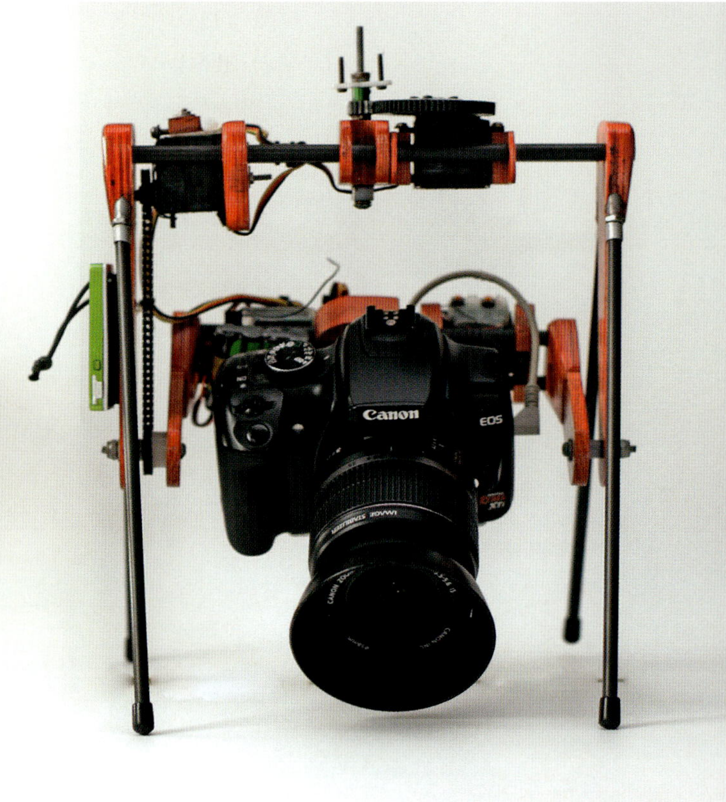

Left: My KAP camera cradles are controlled using radio systems from the model aircraft world. This modified Spektrum 2.4 GHz transmitter can be held in a single hand and has wheels and buttons to aim the camera, lock exposure, and fire the shutter.

Right: The KAP camera cradle I have been using since 2007 carries a Canon dSLR with a wide-angle lens. The apparatus, used to take the majority of images in this book, weighs 3.5 pounds. An on-board GPS logger provides position data for geotagging the aerial images.

from the ground, using transmitters adopted from the model aircraft world. If the great advantage of the AutoKAP approach is simplicity, then the counterpart for radio-controlled rigs is, well, control. This is the method I use. I position the camera overhead by walking around and letting out or retrieving kite line. I aim the camera and fire its shutter, using the radio while I stay at the ground end of the kite line. My handheld radio allows me to rotate the camera through the compass, tilt from horizon to nadir, and change from portrait to landscape format. I greatly value the capacity to control the exact moment of exposure. If the rig is swinging, don't shoot. If vagaries of the wind pull you to the left, don't shoot. With a radio you can wait for your moment. You also have a high degree of compositional control, as each image is aimed and composed independently. I compose by watching the camera rig, sometimes with binoculars though usually with the naked eye, and imagining what it will "see." I generally go about this with preconceived notions of the images I would like to capture, a mental list of the shots I think the scene will offer. More often than not, the mental list is developed iteratively over multiple visits to a photo site. I also use the position of the radio transmitter's controls for feedback about camera orientation, particularly with tilt. The whole process entertains me greatly.

Over the years I have built a dozen radio-controlled camera cradles. Each of these is constructed from the camera outward, balancing the various axes of rotation while installing motors and electronics. The series began with point-and-shoot 35mm film cameras, progressed to single-lens reflex film models with wide-angle lenses, and then shifted to digital cameras when the prospect of rapid feedback proved overwhelmingly seductive. After using a series of compact digital cameras I shifted to a heavier, and more capable, rig designed around the Canon Digital Rebel single-lens reflex. This is the cradle used for most of my South Bay photographs.

My current KAP rig, holding a Canon T2i camera with a 10-22mm wide-angle zoom, weighs just over three pounds. It is connected to the supporting kite line using a Picavet suspension, a small cross with precision pulleys at each end connected by a web of line to two attachment points on the kite line. The mass of the camera cradle causes the Picavet, in gimbal-like fashion, to level itself to the horizon even as the kite

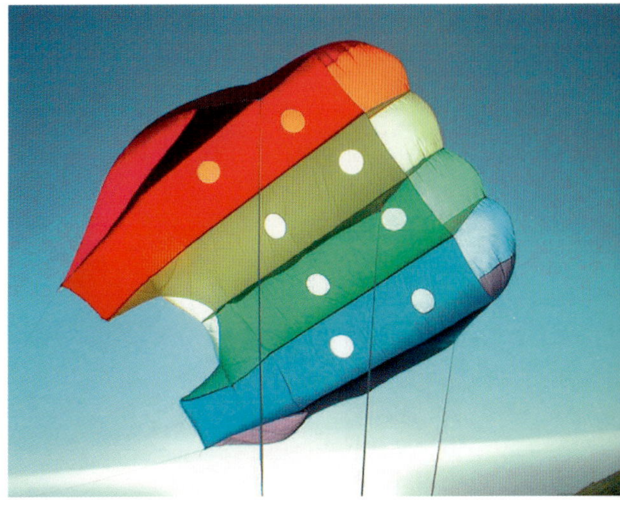

Tough, air-inflated kites like this Sutton Flowform 30 are useful for stronger winds. Instead of using rigid spars, these soft kites hold their shape by inflating with wind-driven air pressure. Simple to launch and fly, they are best suited to smooth winds above 10 mph.

line changes angle. It is a neat trick. The radio has switches and potentiometers that allow one hand to control the camera's orientation and shutter while my other hand flies the kite.

I started KAP knowing precious little about contemporary kites. In retrospect they parallel much that I appreciate in architecture, for the ideal kite, at least for lifting cameras, has a beauty born of function and balance. Modern kites achieve this grace through the clever application of a highly refined material palette including lightweight rip-stop nylon, carbon fiber–reinforced frames, and braided Dacron line. I look back at the paper kites of my youth and shudder. For lifting cameras we desire kites that are calm and stable. They should seek out a spot in the sky and stay there. As a practical matter every kite has a wind range with the low end established by its capacity to lift the payload and the upper end by strength and stability. For very low winds I fly large, gossamer-like kites with rigid frames of improbably thin carbon fiber tubes. Strong winds call for a small, tough-as-an-alley-cat, inflated kite stabilized by a thirty-foot fuzzy tail. My field kit generally includes seven or so kites. With a range of kites available, the goal is to select one that will provide sufficient pull to lift the camera but not much beyond that. This keeps vibration to a minimum and makes hand holding the kite line practical. On some days, my one-kite days, the wind is stable and the selected kite performs well—pure bliss. Other days, with variable winds and poor kite selection, are six-kite days—not so blissful.

As I put my KAP gear to use in the skill-building phase of photographing hither and yon, I had occasion to reflect on my motivations for kite aerial photography. Among the joys of contemporary KAP are the opportunities for invention, the physical challenge of positioning kite and rig, the unusual "once removed" aspect of composition in absentia, contact with a fine group of KAP colleagues, and the distinct pleasure of messing around with kites. My work evolved into a process that I describe as "interrogating the landscape." While it is possible to broadcast video from the kite to the ground as a form of electronic viewfinder, and I have developed rigs to do so, I never use the technique. Instead I prefer to think through the image while standing on the ground imagining what

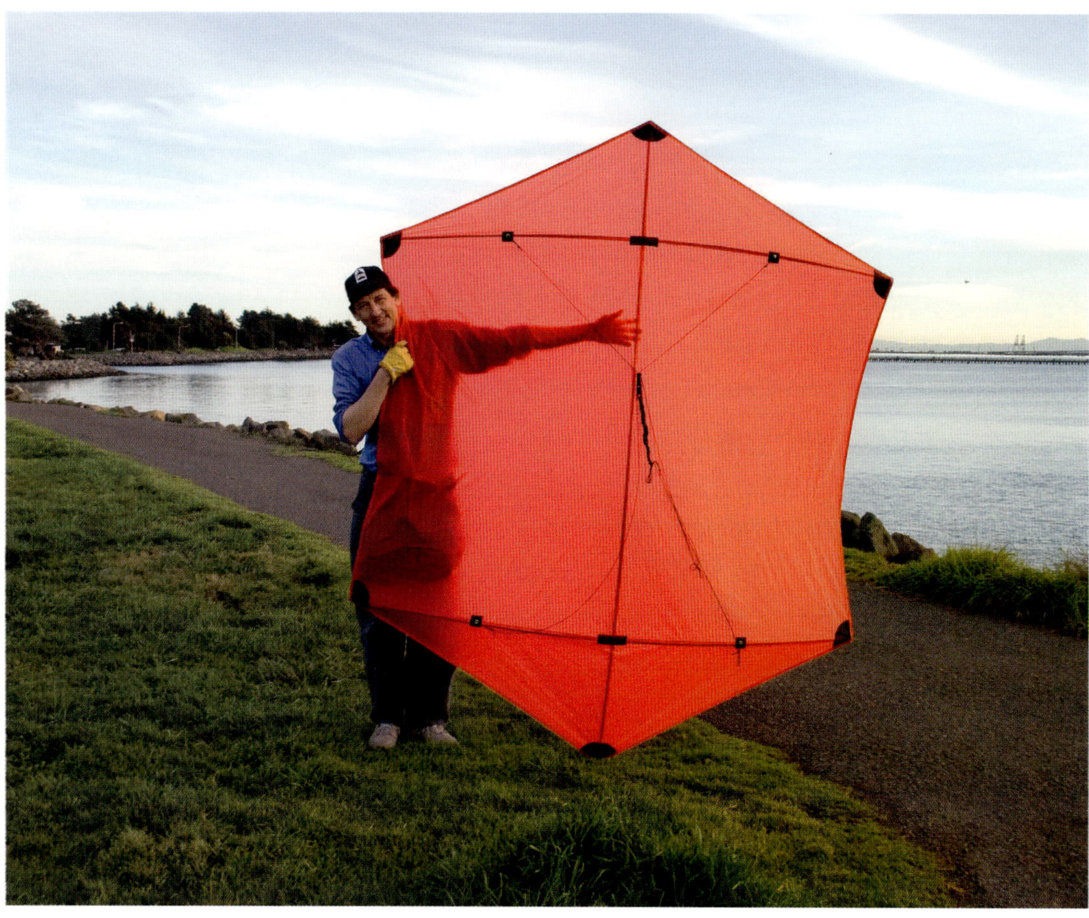

Framed kites like the Rokkaku, based on traditional Japanese geometry, are ideal for lifting cameras in lower wind velocities. This large, eight-foot version is framed with improbably small, carbon fiber-reinforced tubes. It can lift my 3.5-pound dSLR camera cradle in 6 mph winds.

could be seen from that lofty perspective. When I later compare my mental image of the landscape—my visual hypothesis—to the photograph captured, there are invariably differences that warrant reflection and an adjustment of my sense of the place. A great learning experience, this process has become a principal motivator. In the end, kite aerial photography allows me to be overtly rational in the technology and scientific understanding required to capture the images, while simultaneously embracing chance and happenstance. The images thus created are entertaining negotiations between technical prowess and the vagaries of nature.

During a sabbatical leave in 2003, I met Dr. Wayne Lanier, a microbiologist working on a project titled "Hiking with a Field Microscope." Wayne and I started hiking together to take photographs at our peculiar scales—me from a kite and Wayne through the microscope. This was great fun as we learned from each other and the sharing of our views. The collaboration evolved into the Hidden Ecologies project under the aegis of San Francisco's Exploratorium, wherein we explored ways to look at places with an emphasis on juxtaposing scales, collecting different points of view, and encouraging the sharing of ideas. We selected several locations around the Bay Area as test sites, including the South Bay, and thus began a colorful adventure.

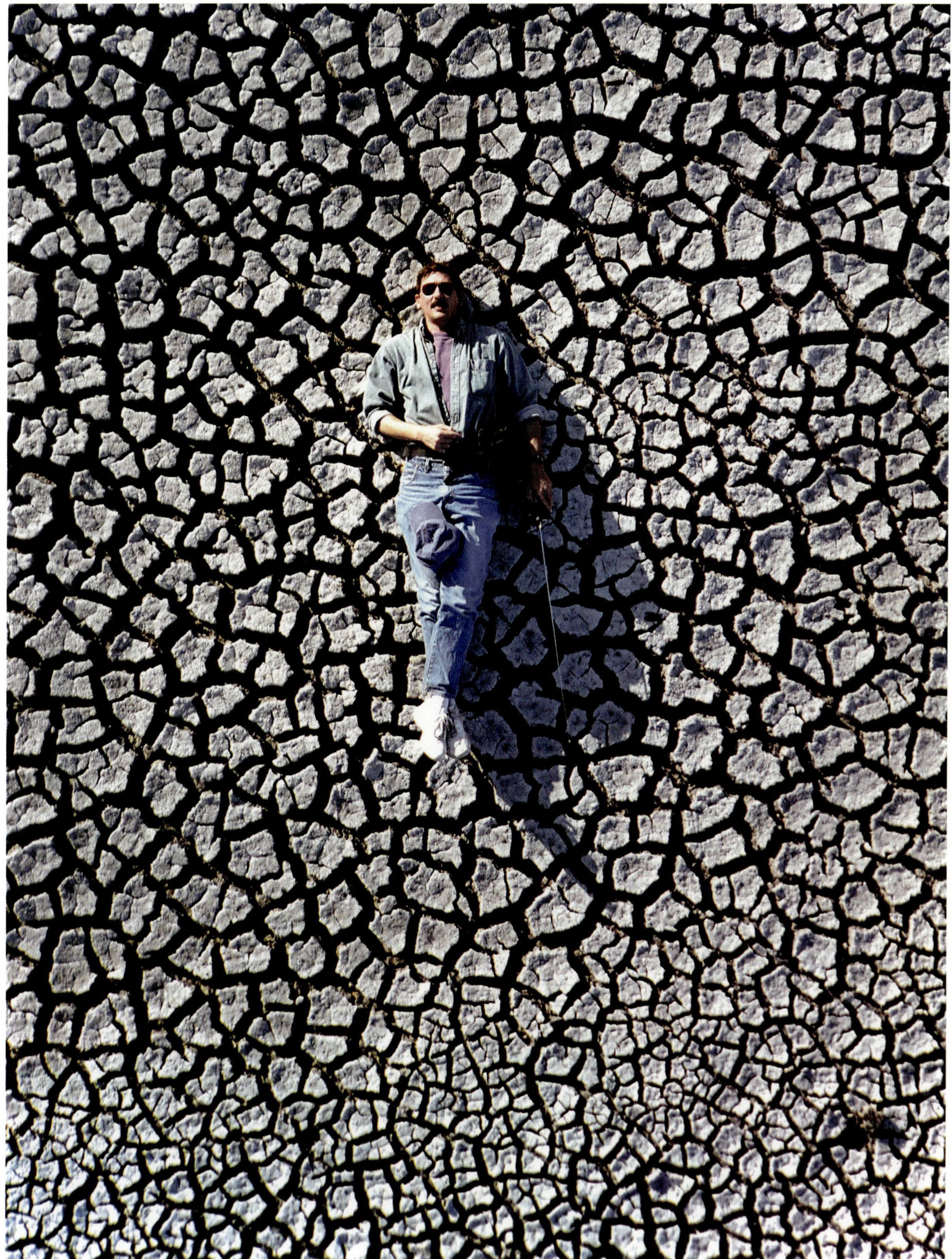

Plate 5. Author on dry pond bottom (Coyote Hills, August 1998)

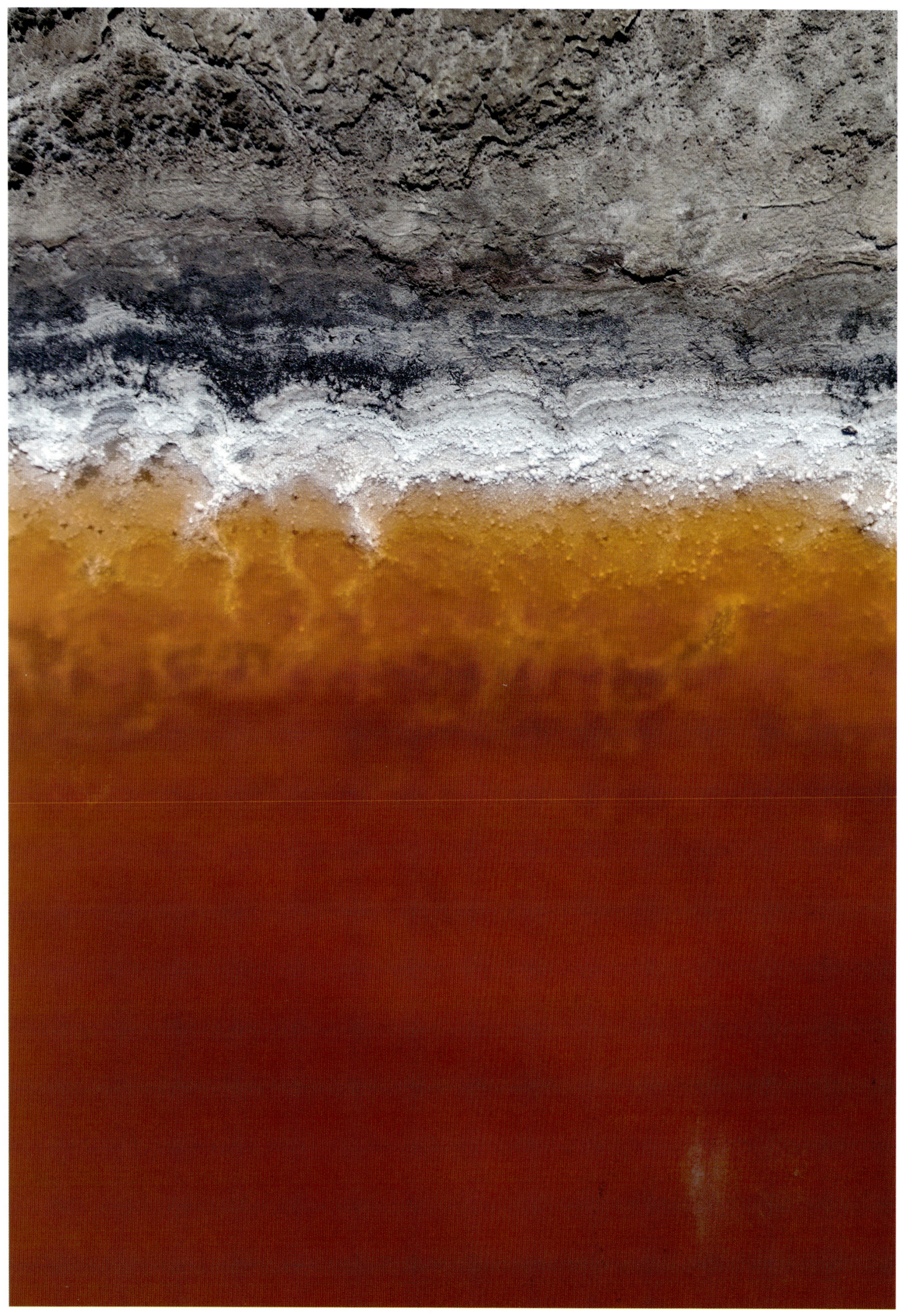

Plate 6. Salt-pond shoreline (Salt Pond E13, August 2011)

2: COLOR AND TEXTURE

HAVING DRIVEN THE PERIPHERY of the South Bay for years, Wayne Lanier and I decided to schedule a proper visit as part of our Hidden Ecologies project. In planning our field trips we imagined the South Bay as worthy of a couple of hikes, a decent prospect for a few photographs and the sampling of microorganisms for Wayne's microscope. For the first visit we picked the Don Edwards San Francisco Bay National Wildlife Refuge Headquarters, located on the southern end of the Coyote Hills in Newark. There, we first approached the salt ponds in a short hike from the east, descending the hill and crossing Newark Slough on a footbridge. The path led to Salt Pond N1 where, at our feet, the pond was a startling red color edged with a lacy, fractal-like shore of white salt. Close inspection revealed that this sanguine soup was alive with brine shrimp.

My impressions of the landscape that day hold true a decade later. Other than the Coyote Hills and landfill operations scattered about, the South Bay is a remarkably flat place—a circumstance offering long vistas to the cities and hills on its periphery. Because of the masking effect of sky reflection from the water's surface, gazing across filled salt ponds, which are quite large, reveals little about pond depth or color. While wildlife is certainly abundant and varied, the place seems isolated, as there are relatively few humans about. But it is a curious and incomplete isolation as aircraft rumble frequently overhead and road traffic glints on the distant freeways. In addition to the salt ponds proper, the landscape holds abandoned pump houses, flow control structures, and other evidence of an industrial past. There is a sense that this place had been used, and used hard.

But the pivotal revelation of that day was the colorful, bizarre stretch of shore at our feet. In this case the pond's vivid color and detail were revealed by our downward angle of view, an overhead prospect also provided by my kite, and therein lay photographic potential sufficient to induce giddiness. This view bundled the promise of abstraction, indeterminate scale, and strong visual fields of color and texture, all ingredients for photographs that could be simultaneously puzzling and pleasing to the eye. After talking to staff at the Don Edwards National Wildlife Refuge and securing permission for a trial run of kite flying, I returned to take several rounds of aerial photographs, an exercise that captured enticing views of that red shoreline. This left me eager to head out seeking other visual delights. Thus began my early engagement of the salt-pond landscape, an enthusiastic quest to capture colors and textures, photographic subjects that played the role of gateway drug in my South Bay immersion.

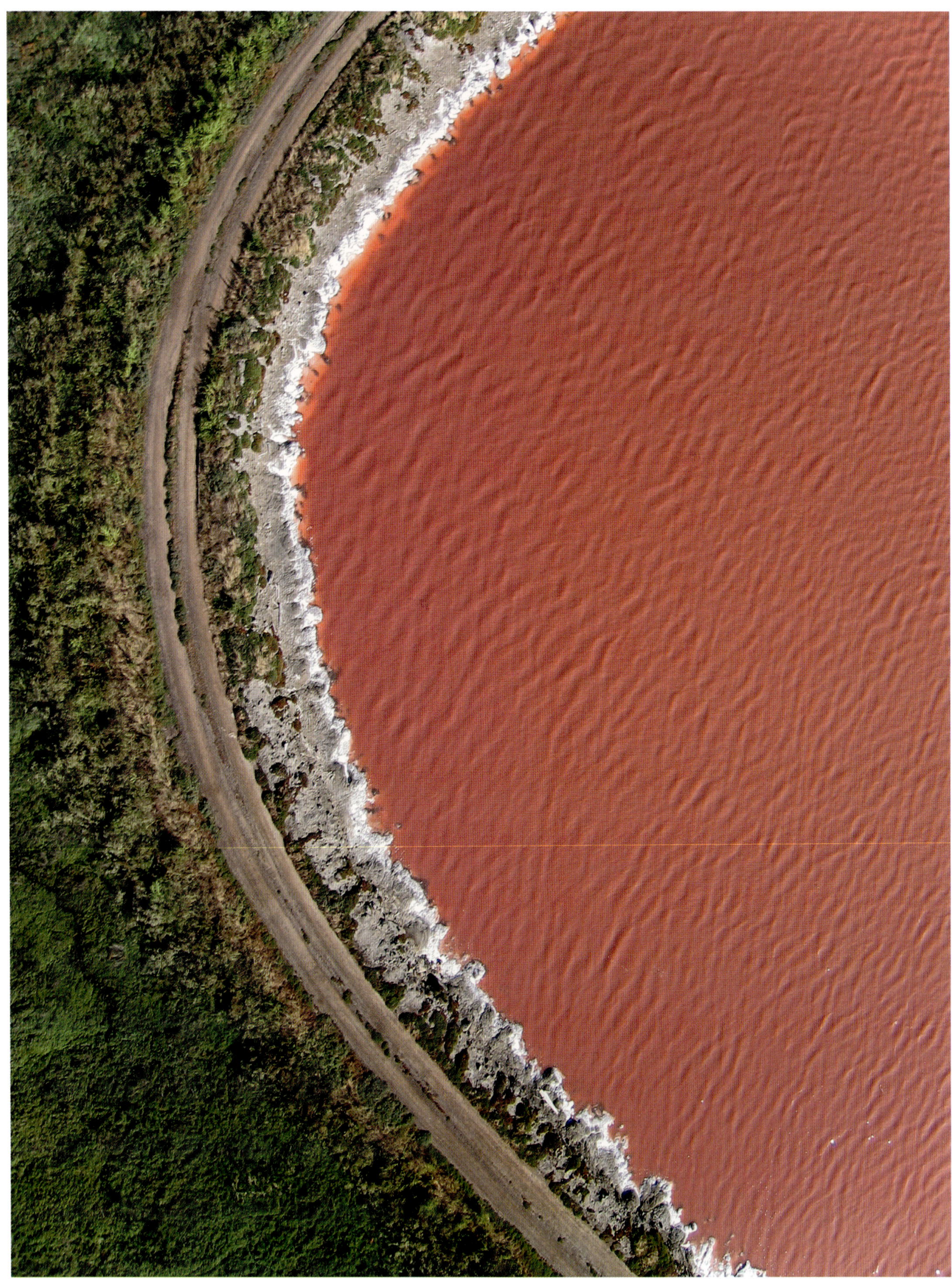

Plate 7. High-salinity pond (Salt Pond M7, May 2005)

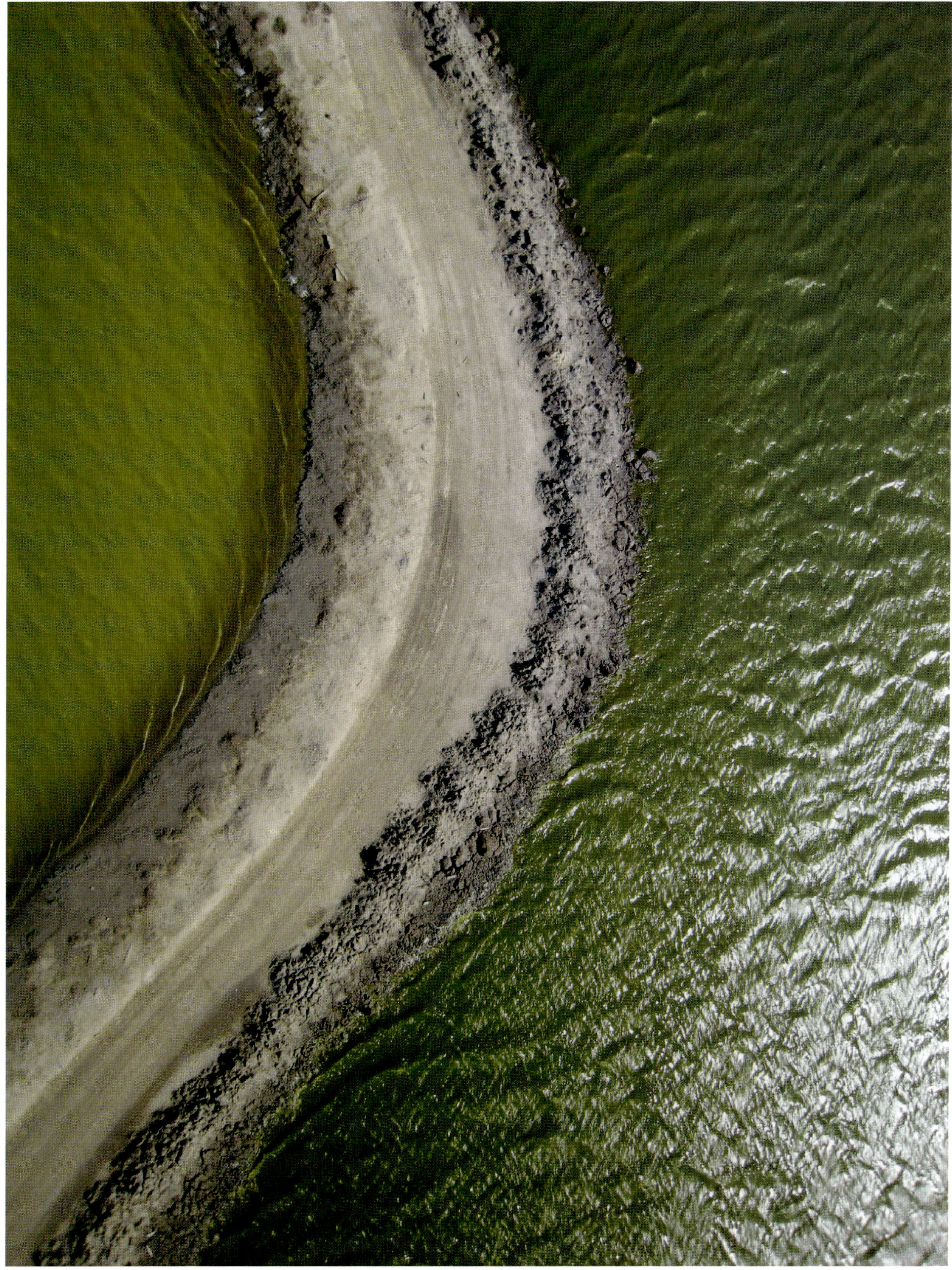

Plate 8. Lower-salinity ponds (Salt Ponds A13 and A15, March 2007)

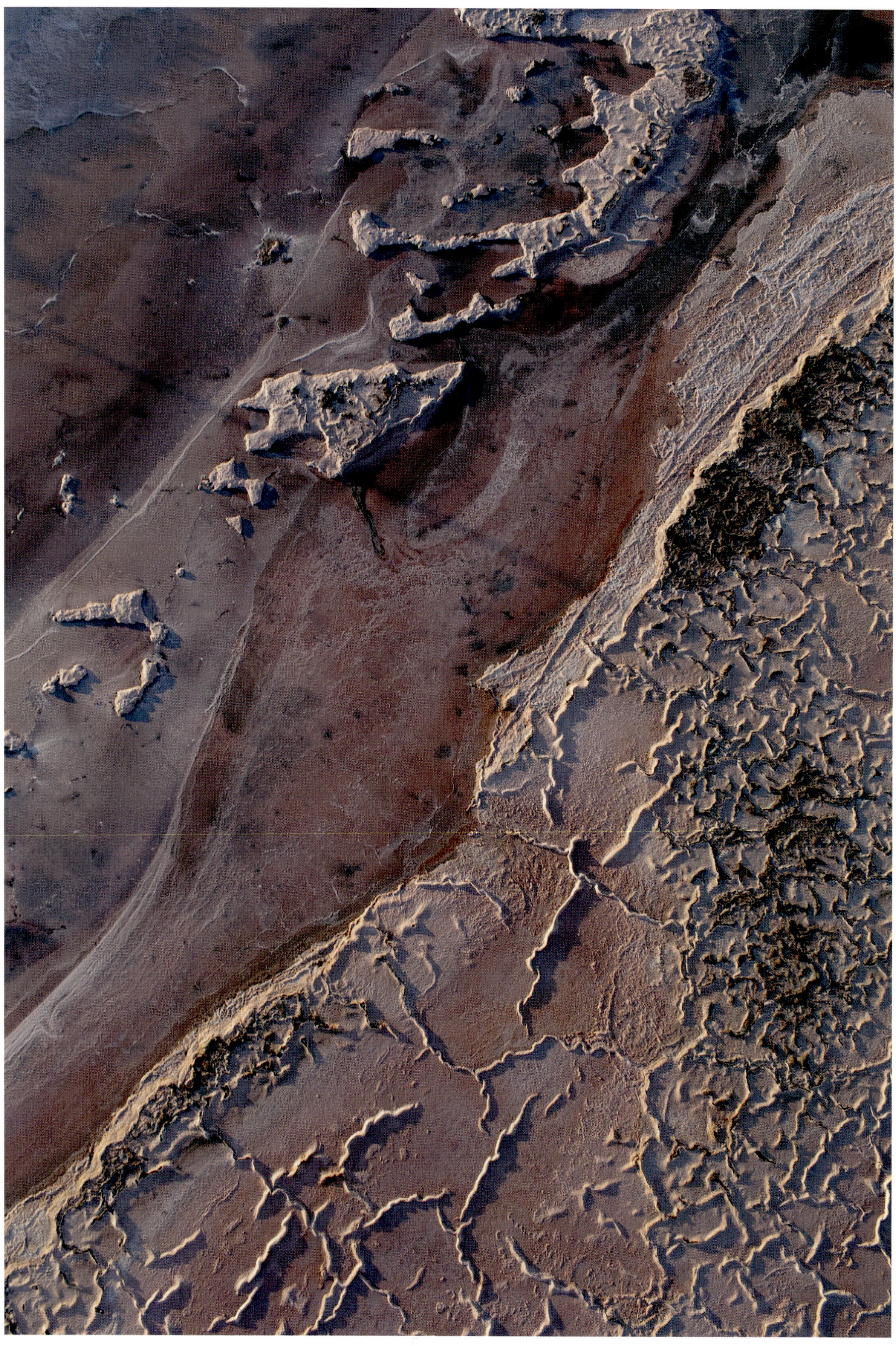

Plate 9. Bottom of inactive pond (Salt Pond A23, September 2009)

Plate 10. My homage to Rothko (Salt Pond E6B, September 2009)

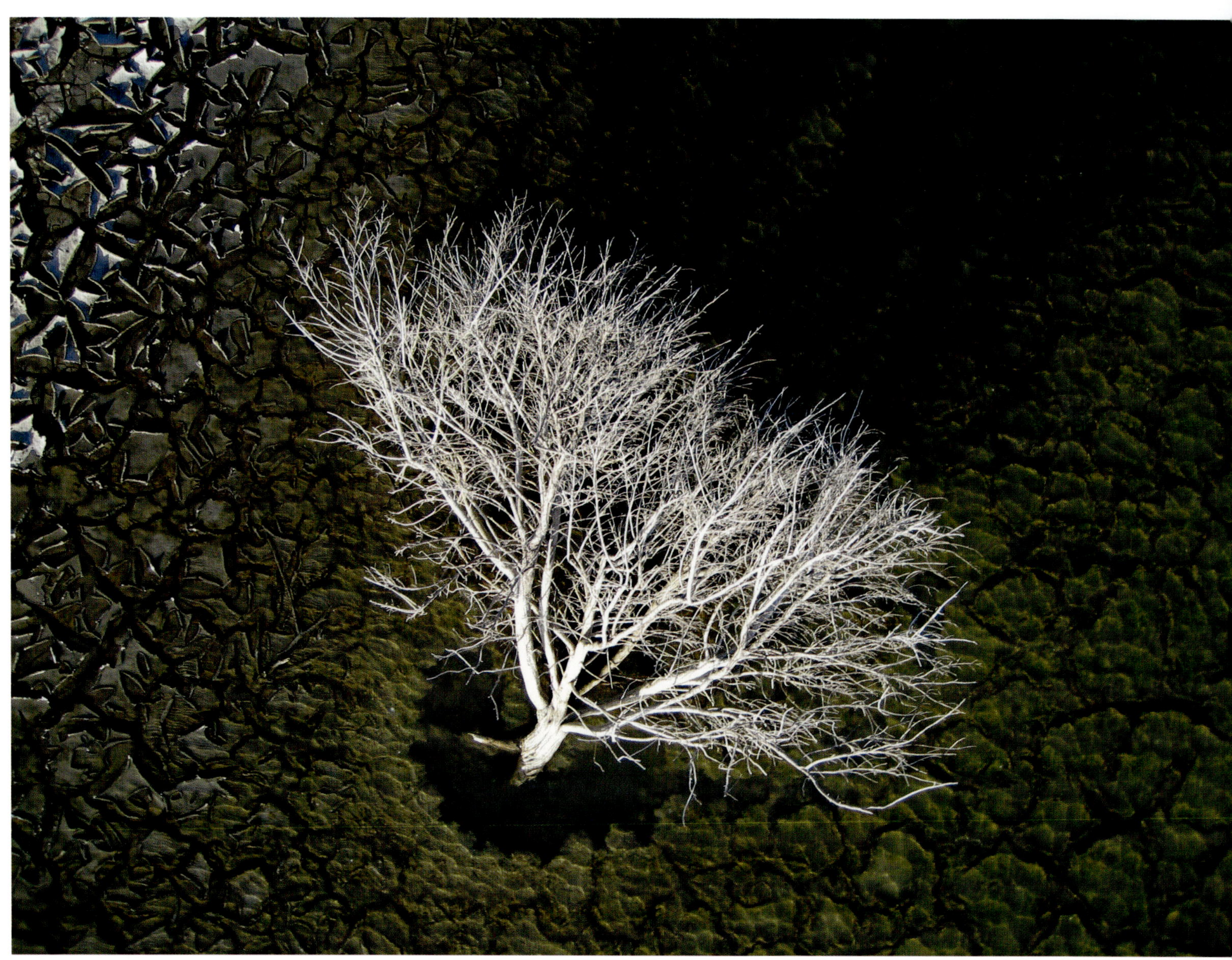

Plate 11. Bush past prime (Salt Pond E2C, September 2005)

As I hiked the South Bay with Wayne Lanier. our views of the same place were simultaneously divergent and complementary. I would look at a place from my kite's perspective to see colors and textures related to physical features like moisture, slope, and salinity as well as signs of previous occupation. Wayne's microscopic view found communities of microorganisms uniquely adapted to immediate environments shaped by the same variables.

It quickly became clear that the landscape's distinctive colors were a visual manifestation of the South Bay's vast salt-making enterprise. Large areas below the San Mateo Bridge are still devoted to the production of salt by evaporation. Rainless summers, a consistent breeze, and abundant sun make the South Bay one of the few places in the United States where salt can be produced by evaporating seawater. The process begins by filling intake ponds with moderately saline bay water. Once marshland, the ponds are now static bodies of quite shallow brine enclosed by low levees built of bay mud. The levees serve to separate the ponds from each other and from the bay proper. As wind and sun reduce an intake pond's volume, minerals remain behind and the remaining brine's salinity increases. At some point the wind, tides, and electric pumps are brought to bear to move the more saline brine to another evaporation pond and the intake pond is filled anew with bay water. This maneuver of shifting brines from one pond to another is repeated often as salinity increases in a patchwork of ponds that once extended over forty-one thousand acres of former marshland. It takes three to five years of evaporation to bring the water from its original 2.5 percent salinity to the 30 percent salinity required to harvest sodium chloride, a trip that may involve ten or more ponds.

The process now culminates in Newark, where Cargill Salt runs the Bay Area's last remaining salt plant. In a series of skillful and delicate transfers salt brine, at this point in the process called "pickle," is taken from the most concentrated of the evaporation ponds and placed in an array of precisely graded crystallization beds surrounding the salt plant. The solution is so saturated that sodium chloride crystallizes and drops, like mineral snowflakes, to the bottom and forms a layer of solid salt on the crystallizer bed floor. The fluid, now called "bittern," is drained off the crystallizer beds before magnesium sulfate is deposited to expose a pond bottom offering five to eight inches of salt available for harvest. This product is crushed and gathered by custom machines and taken to large piles at the salt plant for washing and storage followed by eventual screening and packaging.

As increasingly saline brine makes its way through the sequence of salt ponds it provides habitat for distinct groups of halophilic (salt-loving) microorganisms. Different halophiles thrive at different salinities and, having little biological competition, prosper in great numbers. The practical

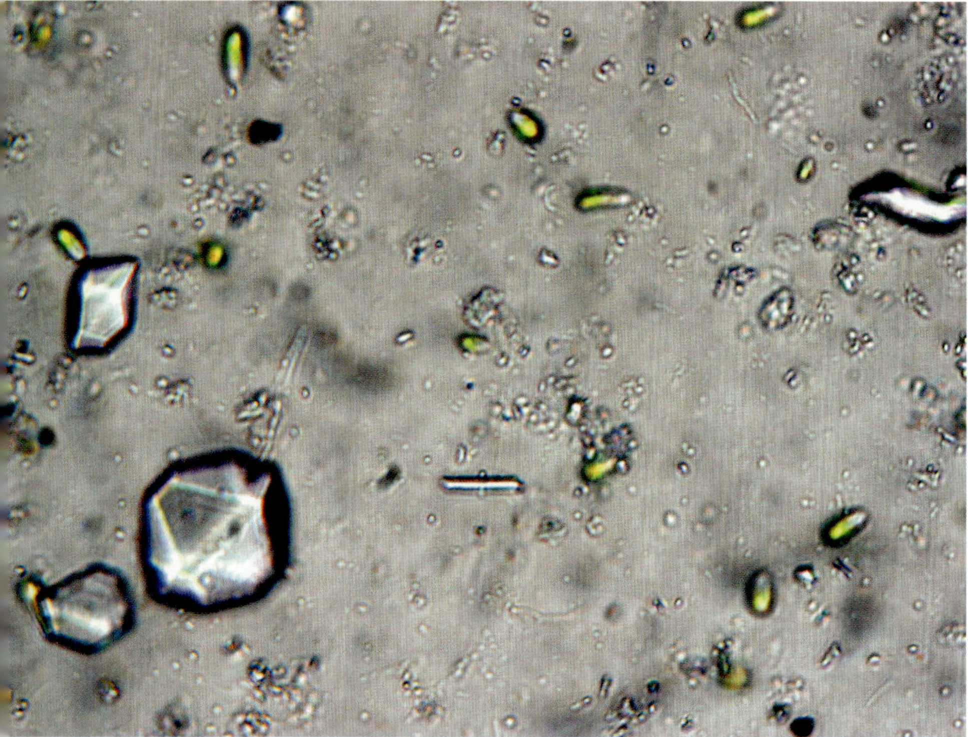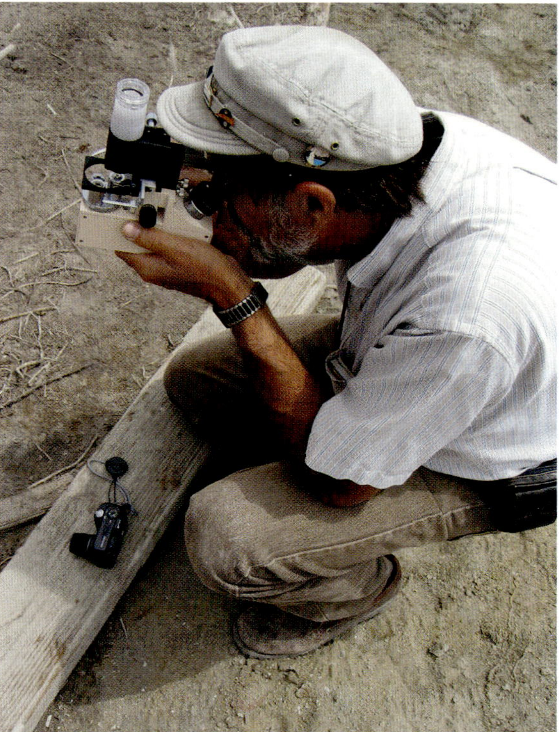

Left: A photomicrograph of the halophilic micro-algae *Dunaliella salina* swimming in water saline enough to have salt crystals.

Right: Wayne Lanier viewing a sample through the dandy Swift FM-30 inverted field microscope, ideal for observations and photomicrography in situ.

consequence is a landscape where color encodes the physical gradient of salinity. Intake ponds, nearly as saline as the bay, are home to many of the species found elsewhere in the bay and tend to be a dull green in color. The medium-salinity evaporation ponds (6 to 16 percent salt) contain mildly salt-tolerant species with Franciscan brine shrimp and *Dunaliella* becoming prominent. These ponds can turn shades of green that are rarely associated with bodies of water, vivid greens with occasional yellowish casts or almost fluorescent tints. As the brine further concentrates, species diversity declines dramatically.

The high-salinity evaporation ponds (18 to 28 percent salt) are populated by biologically ingenious halophiles adapted for survival in salinities where little else can live. *Dunaliella*, the dinoflagellate algae that is green in the medium-salinity ponds, produces a red pigment in highly saline ponds. Brine shrimp proliferate and are joined by halophilic bacteria with red and purplish tints. Pond bottoms in the higher salinities take on an odd buckled texture as gypsum precipitates from the brine and accumulates on the bottom. When viewed from above, these high-salinity ponds begin with a range of brown, orange, and red tints and end with overtones of pink and purple. The most saline ponds, those holding pickle and bittern, host only archaebacteria, extreme halophiles that date back to the beginnings of life on earth. These specialized microorganisms tint the crystallizer beds bright shades of red and purple due to an alternative process for photosynthesis involving large quantities of red-orange carotenoid pigments and bacteriorhodopsin, a process that is a very, very distant cousin to the biochemistry of our retinas.

While combing the South Bay for abstract views of color and texture I slowly built an awareness of the local geography, the workings of the salt industry, and the transitional nature of the place.

After starting around the Coyote Hills my range expanded southward to the south edge of the bay amid the industrial reaches of Sunnyvale. The San Mateo Bridge and its surrounding remnants of nineteenth-century saltworks served as a northern boundary for my efforts. On the eastern side of the bay I poked around the various threshold conditions between the I-880 corridor and the wetlands proper, a zone that was relatively accessible for kite flying. The western side was less hospitable, as a series of small airports created zones where kite flying was problematic.

My weekends were often spent within these boundaries, hiking new areas and bagging, like trophy animals, new colors in the salt-pond sequence. My nodal landmarks expanded to include trailheads and access points for the South Bay, including the Eden Landing Ecological Reserve, Dumbarton Point, Alviso, Ravenswood, and Bayfront Park. I also began to sort out patterns of infrastructure festooning the South Bay landscape—high-tension power lines (yikes), water pipelines, landfills, rail lines, bridges, and roadways—spending many happy hours with old maps and survey photographs to develop a sense of what appeared when. But as my wanderings expanded, my mental map of the place became increasingly confused as I tried to place another of the area's major influences. This was the Don Edwards National Wildlife Refuge and associated programs, entities that had acquired large tracts of the South Bay with an agenda of managing the landscape for wildlife habitat, flood control, and outdoor recreation.

In 1972 Congress passed, and President Nixon signed, a bill authorizing establishment of what was to become the Don Edwards San Francisco Bay National Wildlife Refuge. Thus began a different chapter of South Bay transitions, a new scenario in which salt production would coexist with management of the land for the public good. The new regime gained traction during the 1970s and

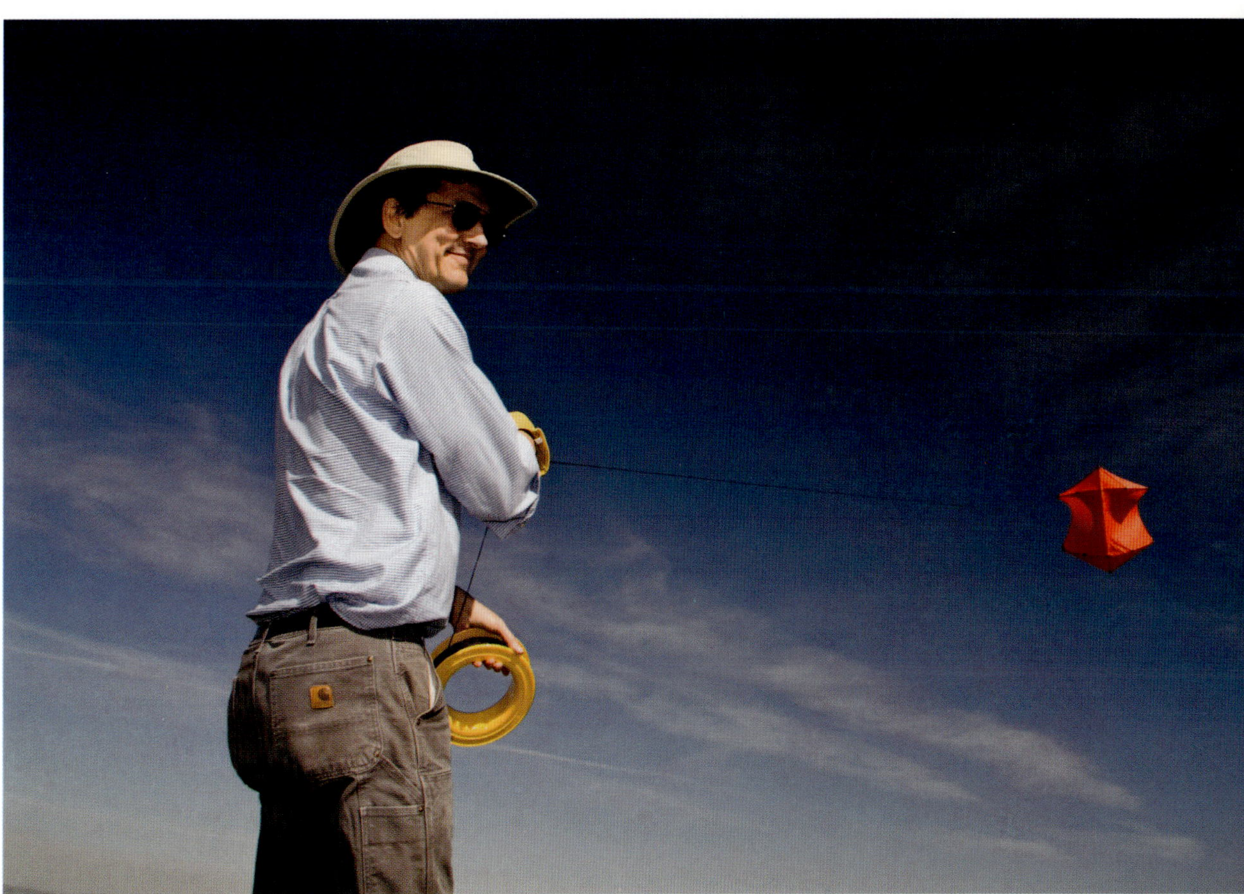

The author launching a smaller Rokkaku. Photograph by Danny Wilson

1980s with the purchase of former wetlands, largely from Cargill Salt, and the formation of alliances that pooled the interests of federal, state, and local governments with private foundations and nature organizations. By 1980, the refuge had acquired title to the majority of the South Bay's current and former wetlands, with Cargill retaining perpetual mineral rights to most of the refuge holdings, which continued to operate as salt-evaporation ponds. The landscape, while protected from future development, remained essentially unchanged. Then, in 2003, a complicated and innovative agreement was struck between Cargill Salt and a coalition built around federal and state agencies. This deal greatly reduced the area of land involved in salt production and marked the beginning of the South Bay Salt Pond Restoration Project, a multi-decade effort to envision and implement revisions to the South Bay landscape driven primarily by the goals of the refuge. With project areas in Alviso, Ravenswood, and Eden Landing, this effort is developing an ambitious portfolio of habitat interventions ranging from marsh restoration to ponds managed for the benefit of specific wildlife species.

In my first forays into the South Bay I naively anticipated that I would find salt ponds and refuge lands existing side by side as discrete and quite different parcels. Instead I came to realize that the refuge and salt industry often shared the same ground in a relationship that acknowledges the valuable and diverse habitats provided by the salt ponds and Cargill Salt's expertise in managing this specialized landscape. In the field I encountered large plots in the South Bay that were not actively involved in salt production. These included small remnants of the original marsh, once an impressive fifty-three thousand acres, and larger areas that seemed to be former salt ponds gone fallow. These fallow ponds had their own visual intrigue. As they filled with the winter rains and evaporated over the summer, residual salt from some previous epoch would dilute and concentrate to produce small pools that followed the color patterns of the large salt ponds. In other locations levees had been breached so that former salt ponds were returned to tidal flow.

With great interest I began to appreciate the South Bay's complexities in jurisdiction and, furthermore, that changes in ownership and management were driving a transformation of the landscape. Indeed, massive change was underway. And then, as I reviewed my early aerial photographs of the area, another and perhaps more intriguing realization developed. The current period of transition was not an isolated circumstance. This landscape has experienced frequent and successive stages of transformation for over one hundred and fifty years. There, in my images of color and texture, one could also see traces of the past, presenting very much like incremental design sketches in layers of architectural tracing paper. In a landscape where the height of features plays out in just a few feet, here were visual patterns linked to previous epochs in transportation, infrastructure, and salt production—sometimes faint, sometimes distinct, occasionally superimposed several layers deep. And below them all were dormant channels from the vast marsh that existed before human intervention.

What began as a photographic romp through a visually compelling landscape slowly shifted toward documenting the landscape's history and deciphering the traces of it that were evident in my aerial photographs. While doing so I became much more aware of the South Bay Salt Pond Restoration Project and ended up getting special use permits from the US Fish and Wildlife Service and the California Department of Fish and Game (now Fish and Wildlife) for kite flying and photography in portions of the South Bay where these activities are normally prohibited. In return I set about documenting the South Bay landscape in earnest as it changed below my camera.

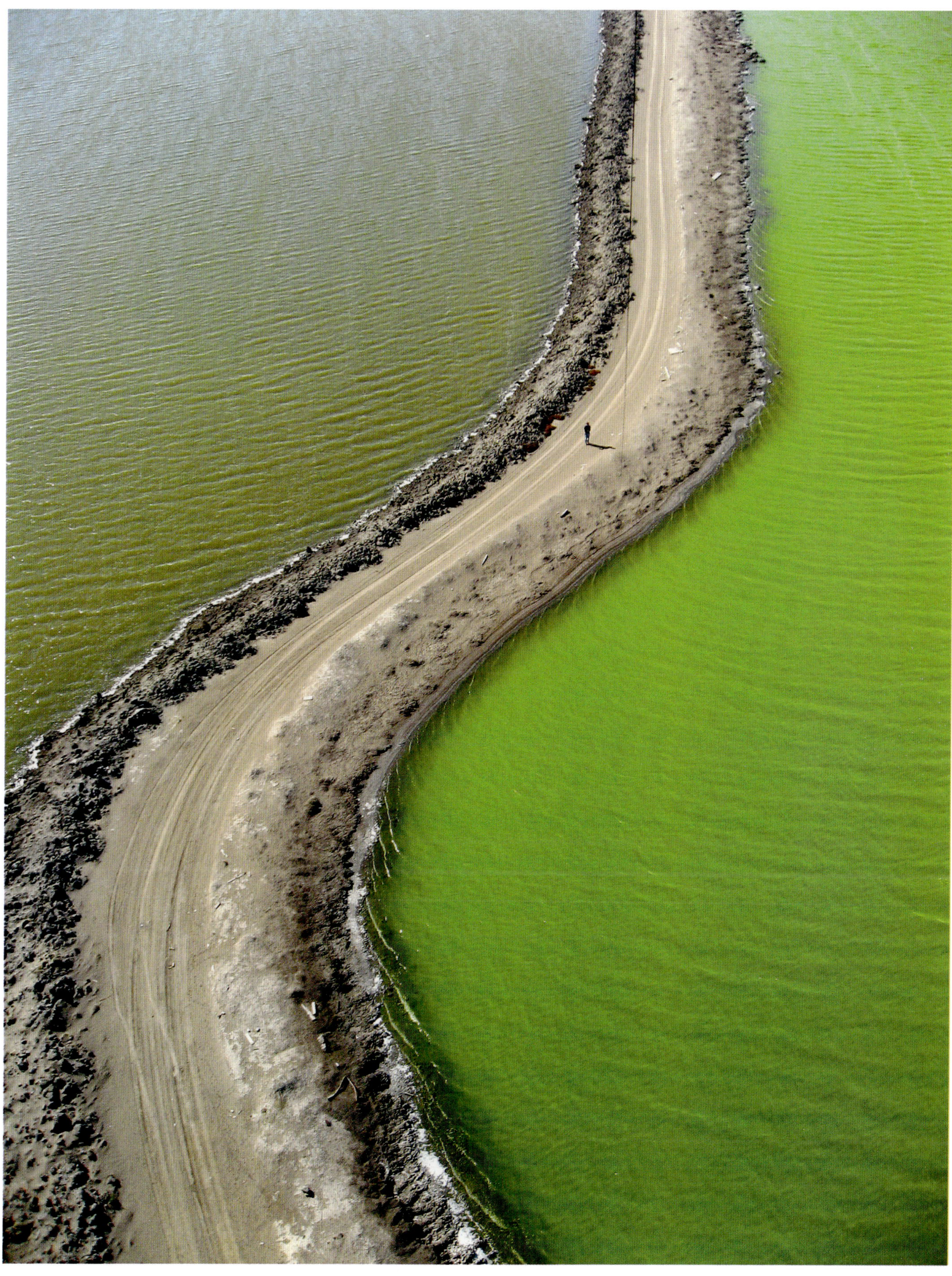

Plate 12. Walking the levee (Salt Ponds A13 and A15, August 2006)

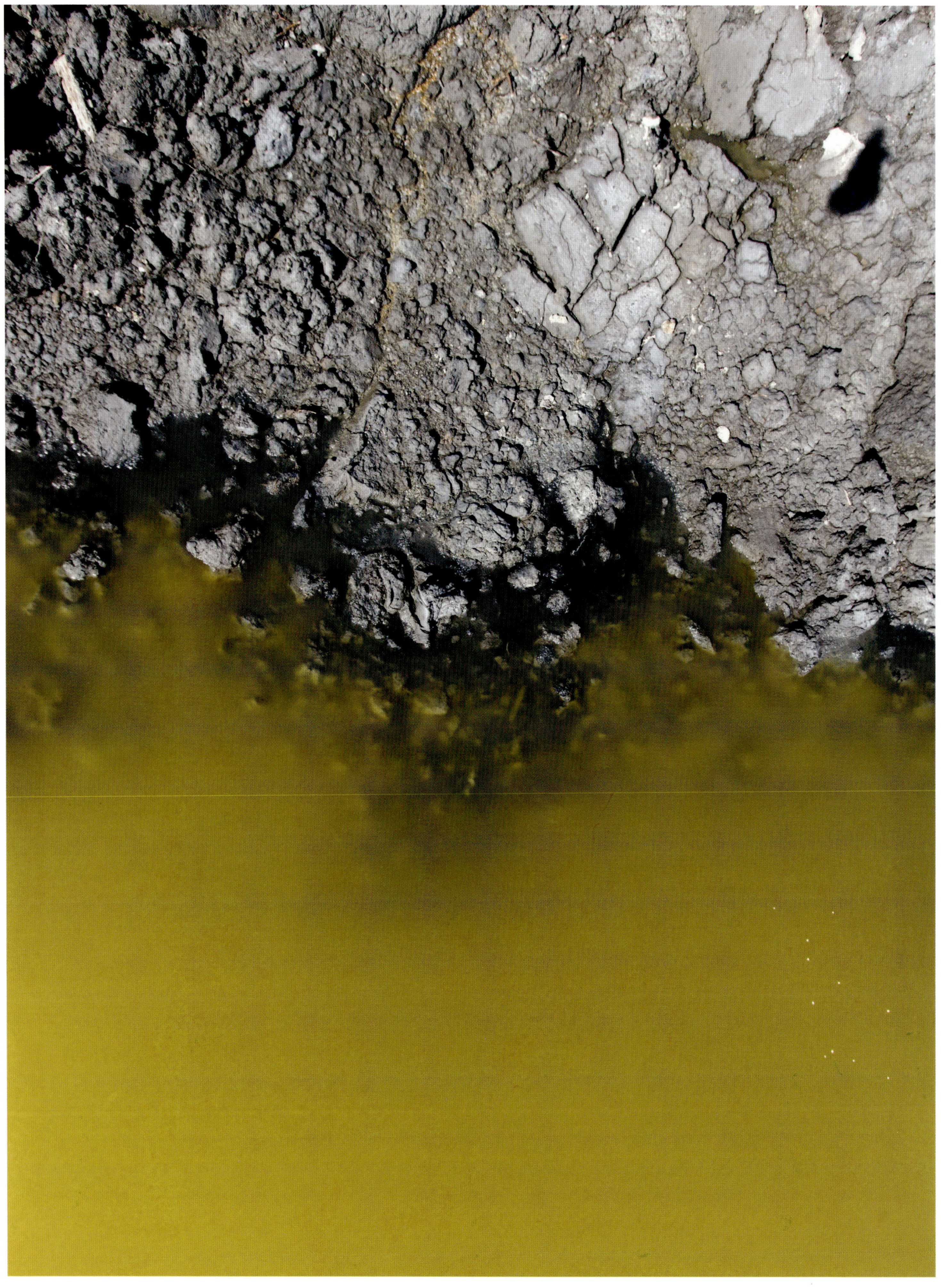

Plate 13. Fresh levee (Salt Pond N2, former course of Beard's Creek, April 2012)

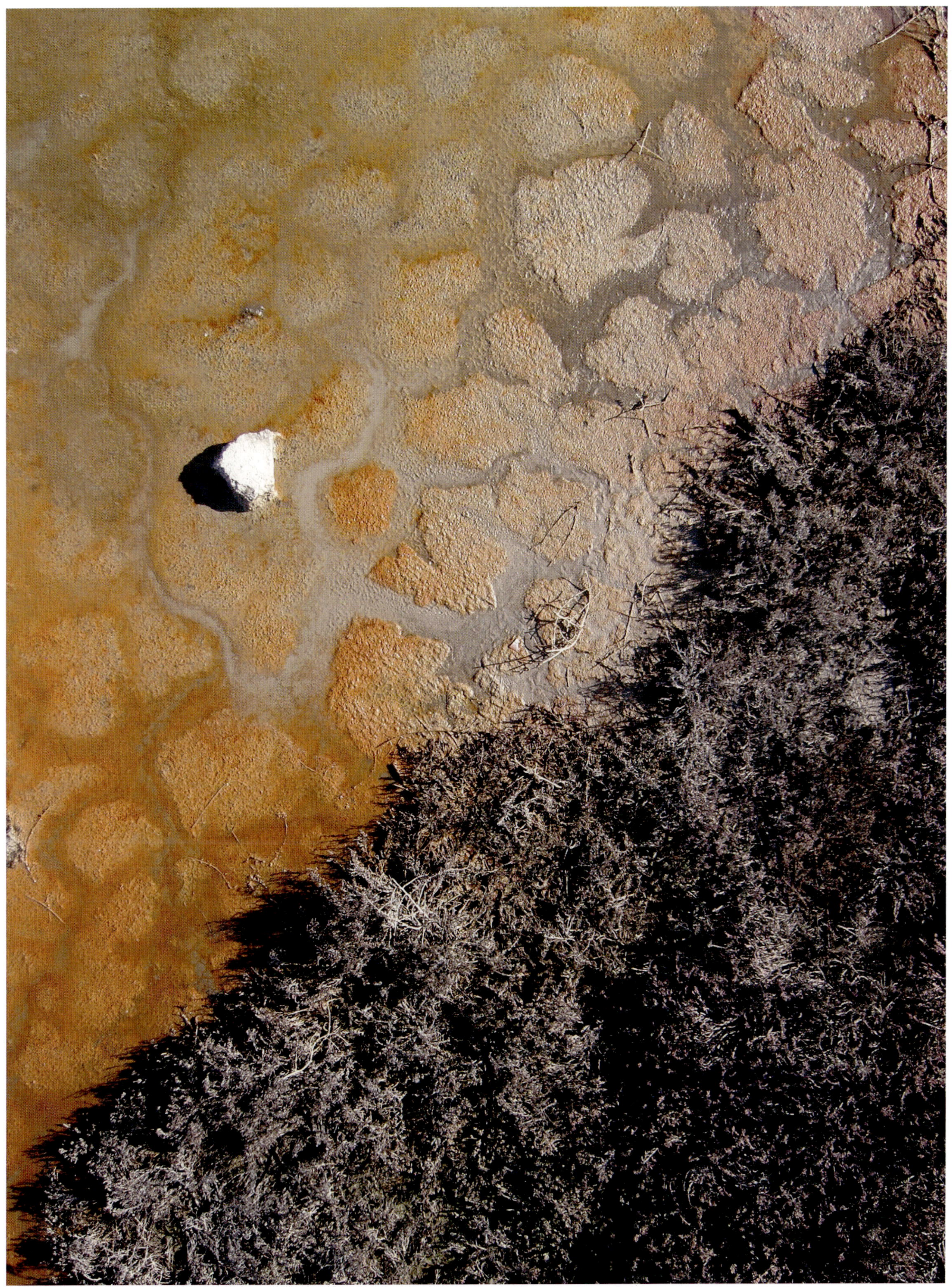

Plate 14. Pickleweed verge (Railroad right-of-way near Alviso, December 2006)

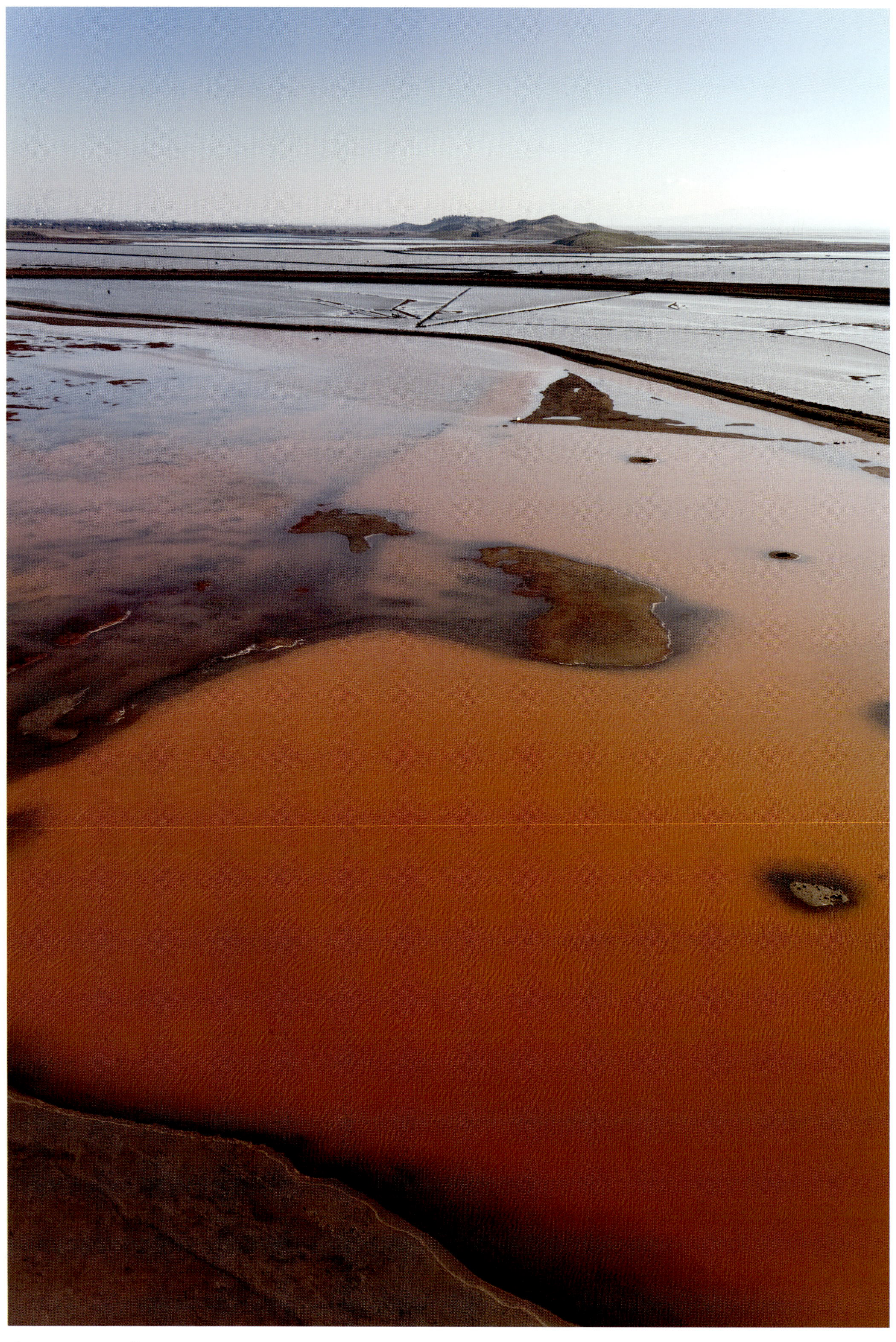

Plate 15. Coyote Hills (Salt Pond E6B, November 2009)

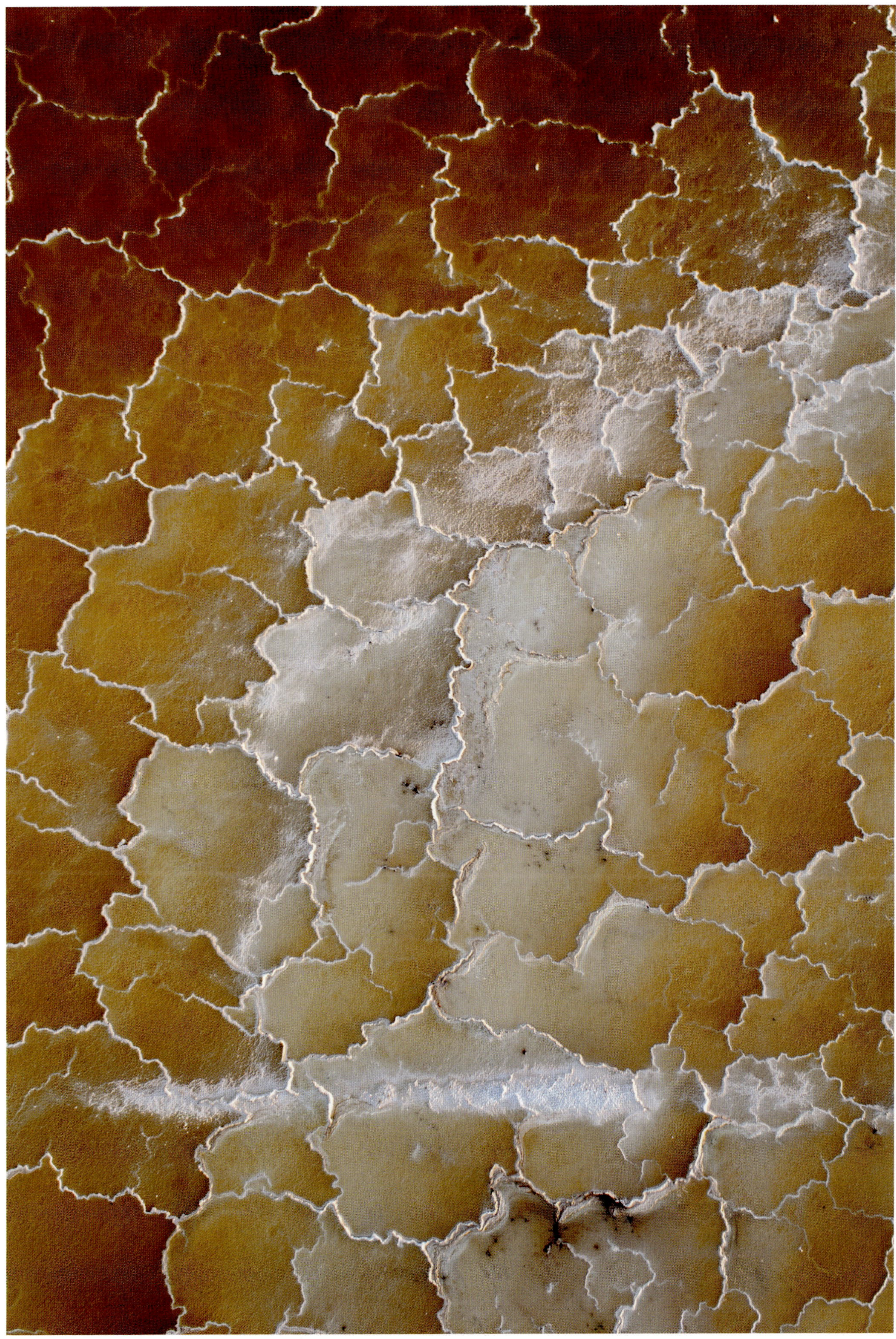

Plate 16. Buckled gypsum (Salt Pond E6B, September 2009)

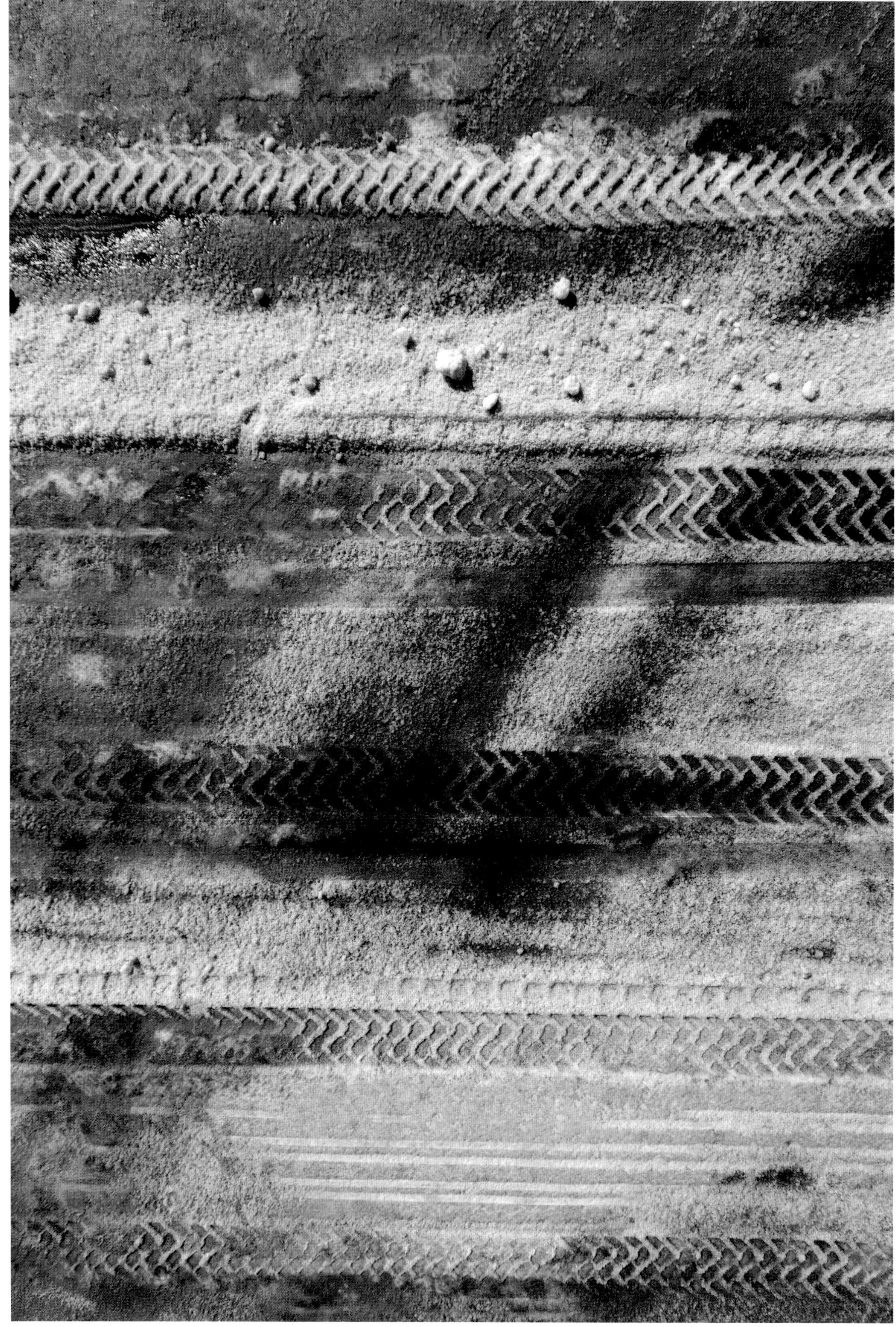

Plate 17. Equipment tracks (Cargill Newark crystallizer beds, October 2010)

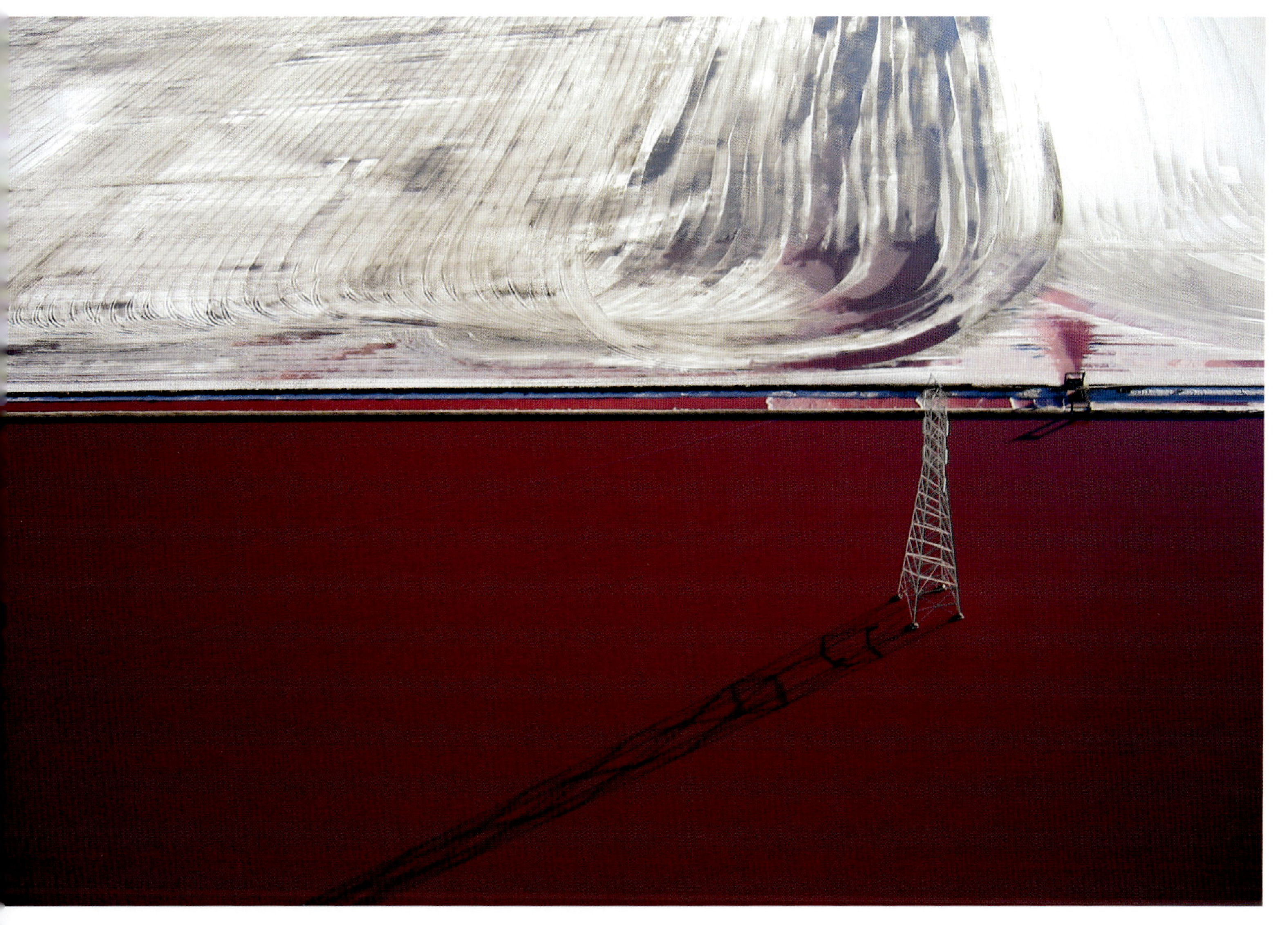

Plate 18. Crystallizer beds (Cargill Newark crystallizer beds, September 2005)

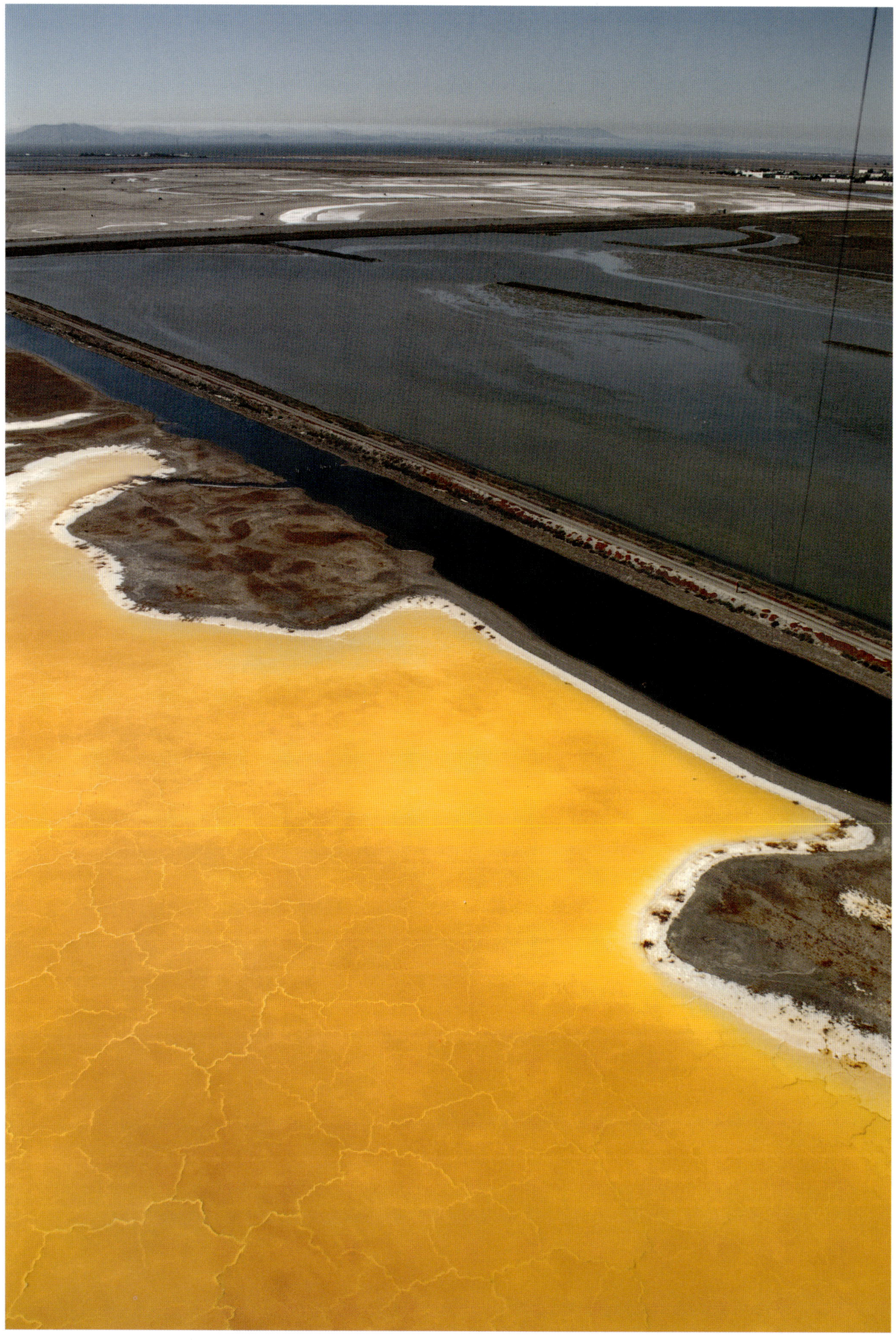

Plate 19. Changing regimes (Salt Pond E6B, September 2009)

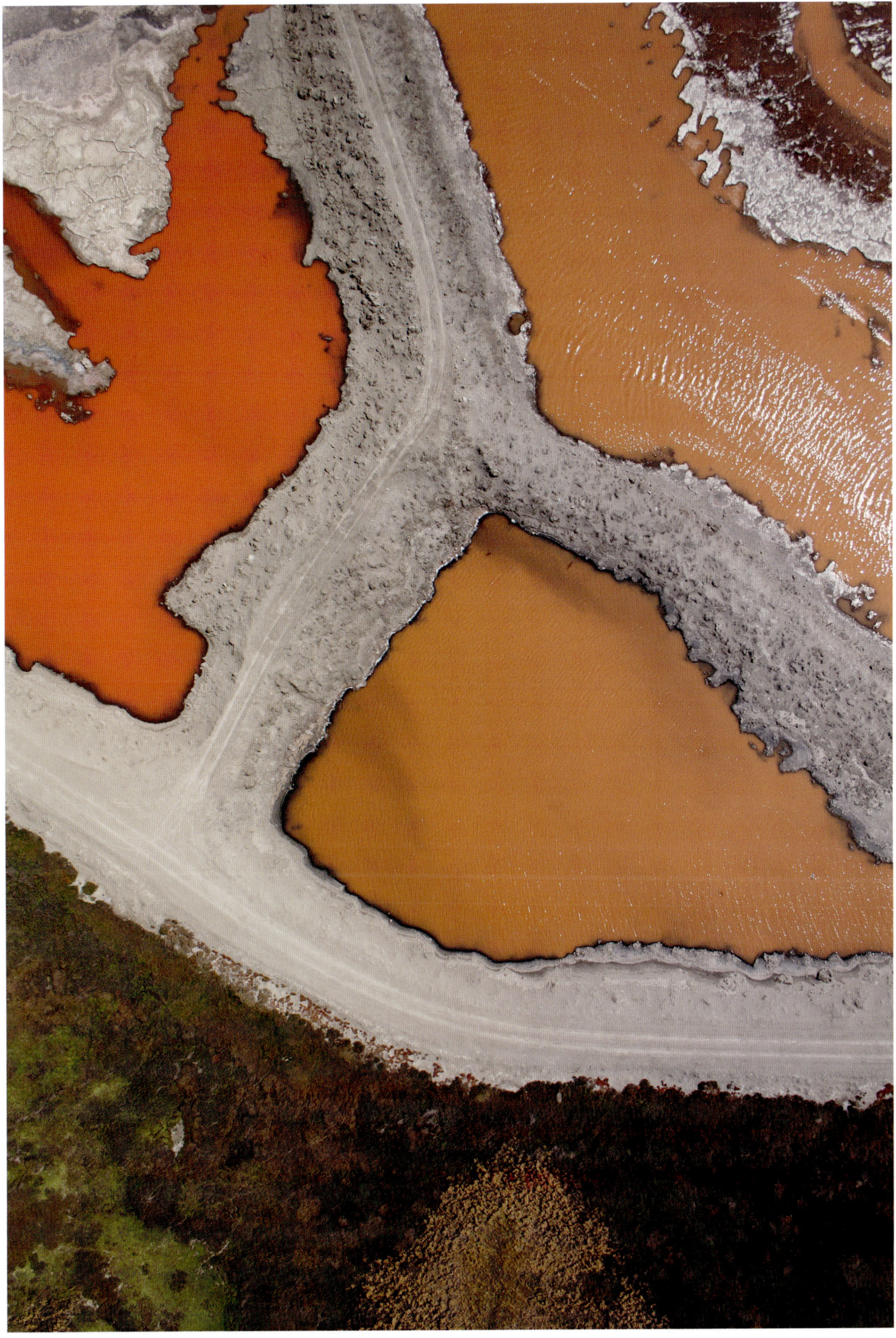

Plate 20. Dredge lock (Salt Pond M4 and M5, July 2008)

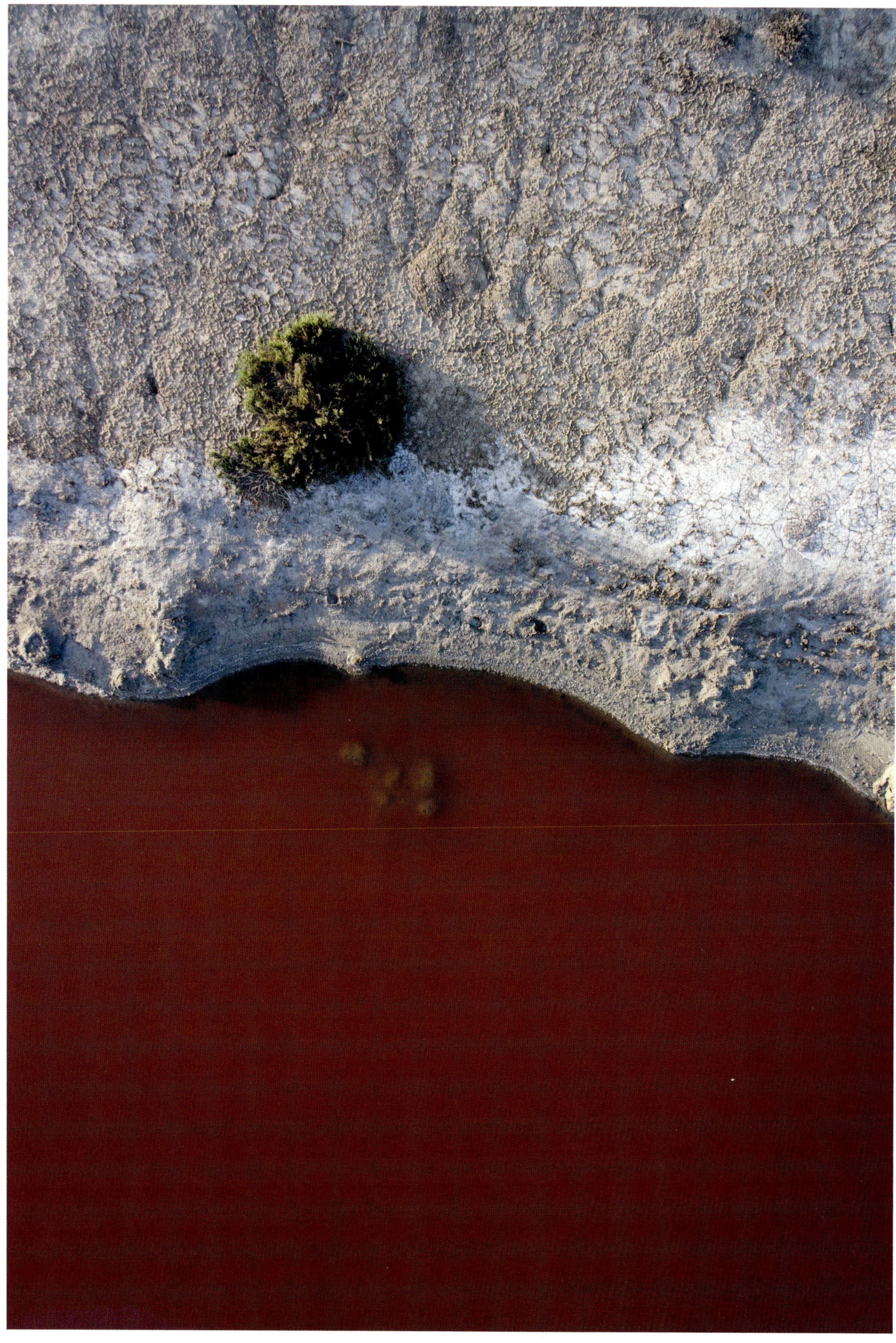

Plate 21. Abandoned borrow ditch (Salt Pond R5, Bayfront Park, May 2012)

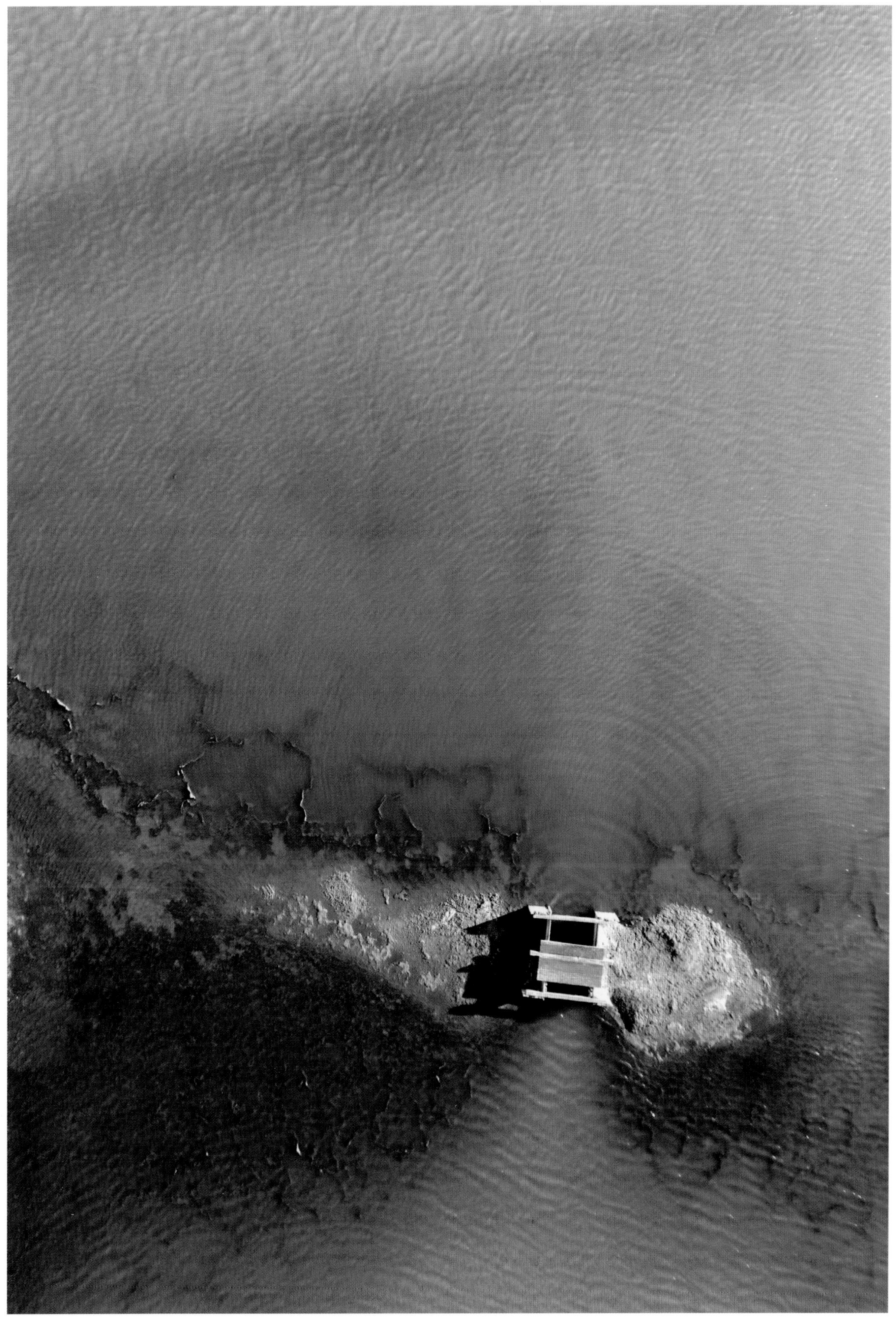

Plate 22. Derelict flow gate (Salt Pond N2, former course of Beard's Creek, April 2012)

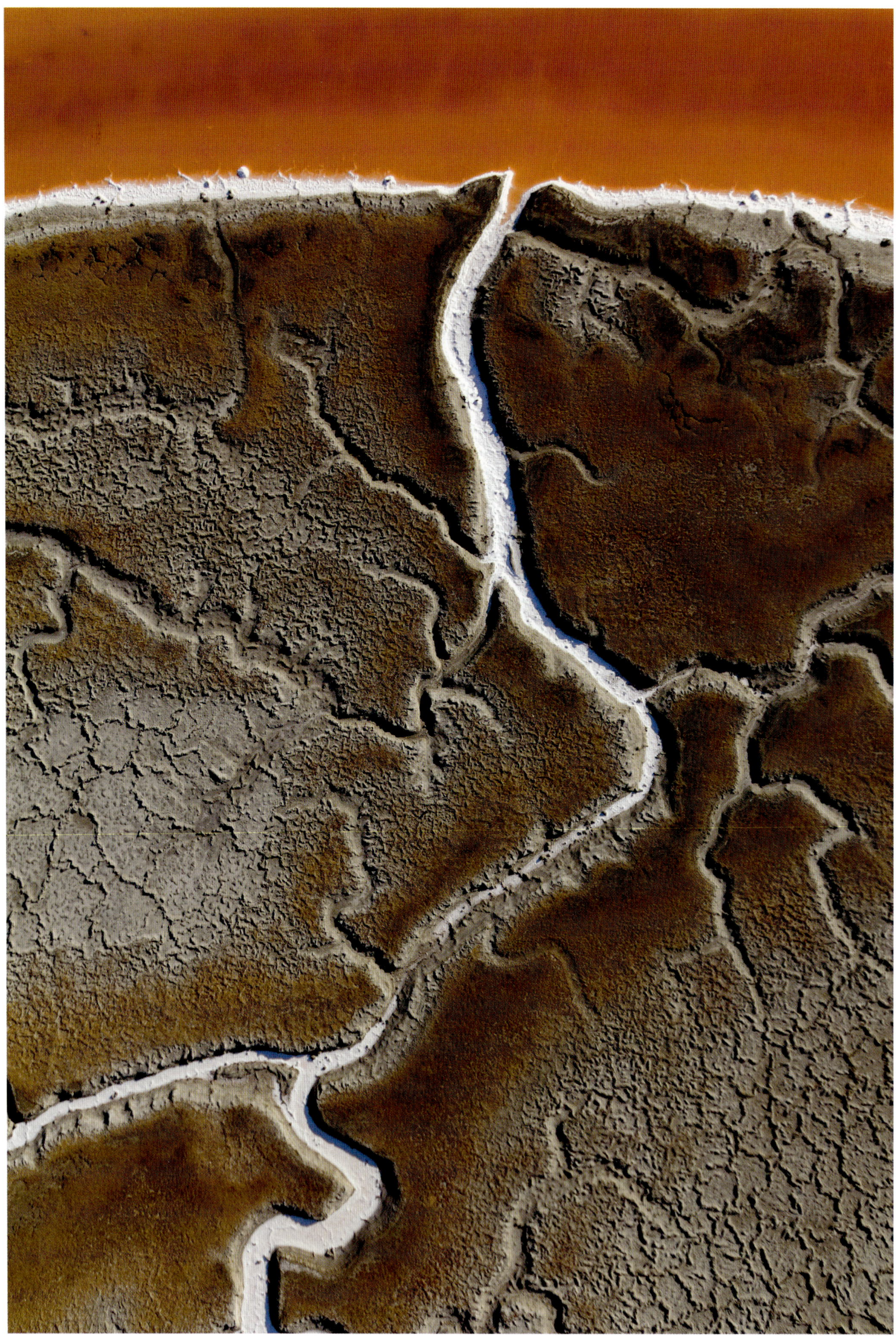

Plate 23. Fallow pond (Salt Pond R5, August 2011)

3: PRIMORDIAL MARSH

When members of Gaspar de Portolà's expedition crested Sweeney Ridge in 1769, they were greeted with sweeping views of San Francisco Bay. To their right was a vast tract of marshland in the South Bay, an area later explored by Juan Bautista de Anza's troop. The South Bay's relatively shallow waters were organized into concentric zones with the bay proper lying at the center. Expanding outward there existed a generous band of tidal mudflats, then an impressively broad plane of marsh, and finally the upland margins, alluvial valleys and creek plains of the adjacent shorelands. Through this watery landscape wandered marsh channels and sloughs totaling three thousand miles in length. Tides animated the sloughs—and the marsh they served—with natural cycles of ebb and flow that created rich habitat for local and wayfaring species alike. It must have been an extraordinary sight.

Relatively little—about one-sixth—remains of the original wetlands, which had maintained equilibrium in the South Bay for approximately two thousand years. Scattered parcels were left here and there in places that were inconvenient to develop, but the rest was lost to human intervention. Filling wetlands for development, think Foster City, destroyed about a quarter of the original marsh, while the creation of salt-evaporation ponds claimed the rest. Diking for salt ponds began in the 1850s around Mount Eden Creek. By the early twentieth century, salt ponds had become the dominant landscape element they are today. Thus, it surprised me to find the familiar sinuous shapes of marsh channels and the round flats of marsh pannes in my aerial images of the salt ponds. It turns out these marsh features, lying dormant, are still very much with us.

While traveling the South Bay, exploring new trails and ponds, I found dormant marsh channels at every turn. Here they are in the surface topography of a dry, fallow pond, and over there as soft-edged depressions in the bottom of an active evaporation pond. With a presence ranging from distinct and unambiguous to the faintest gestural remnant, these "antecedent channels," as the restoration folks call them, serve as testimony to a landscape that existed just a few generations ago. I confess to somber moments when considering what little remains of a once magnificent wetland. My principal comforts are a realization that matters were much worse around 1960 and the sense that environmental efforts since that nadir have helped turn the corner. What was disturbed by a century of active development might well be repaired over a similar time frame. Meanwhile, the familiar shapes and patterns of the marsh channels sit patiently, as though waiting for the tides to return.

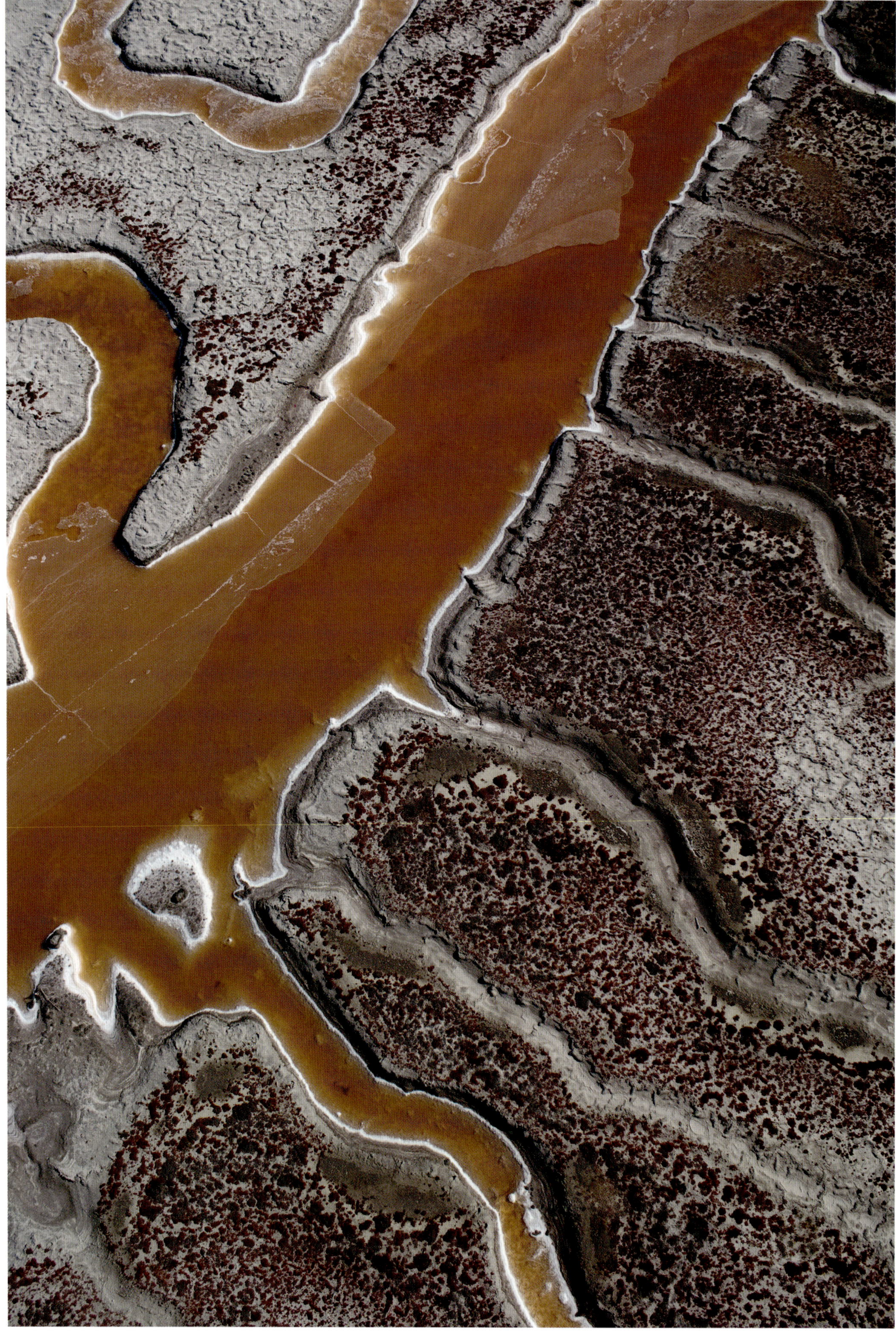

Plate 24. Marsh channels awaiting the next chapter (Salt Pond R5, July 2009)

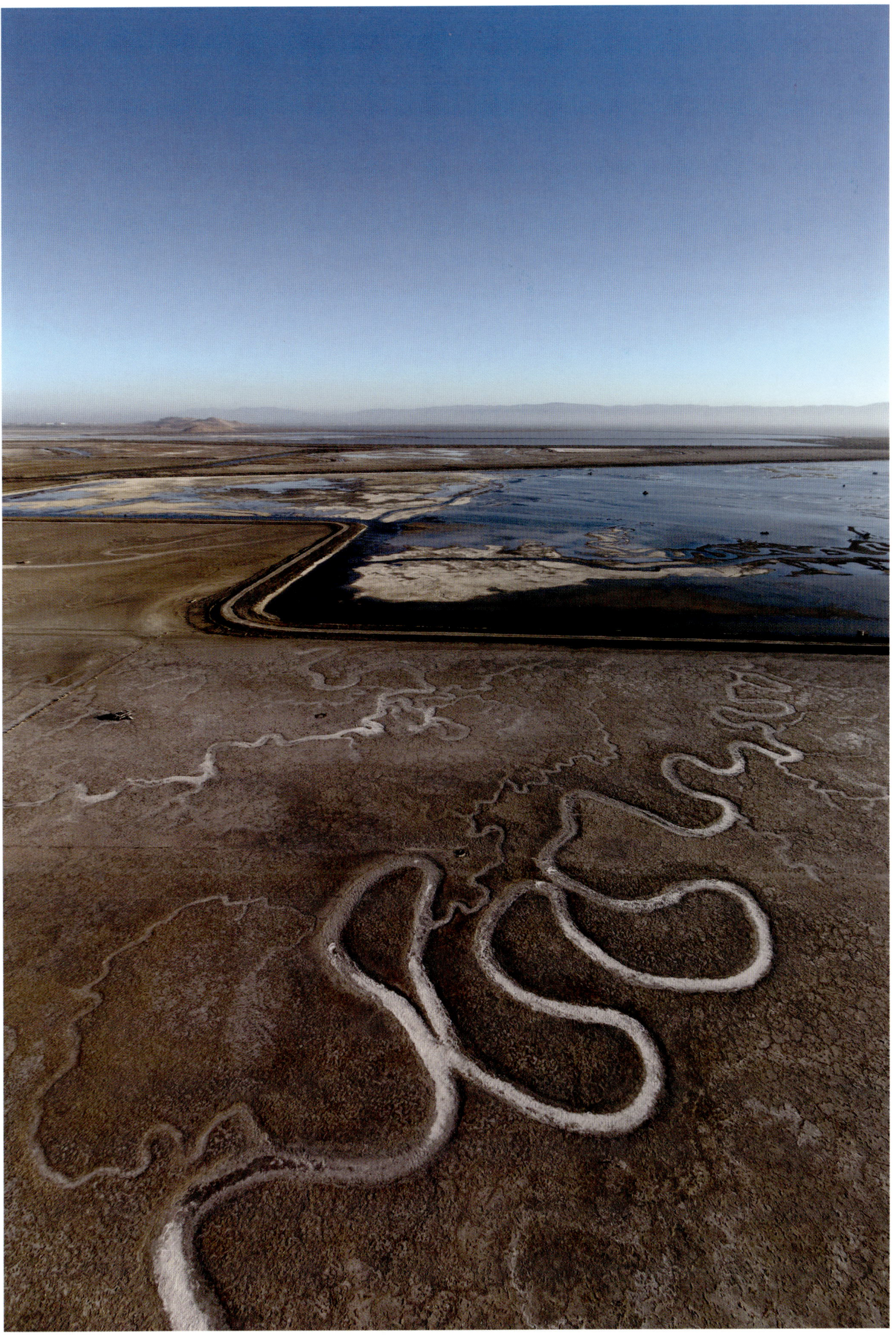

Plate 25. Traces of long-dormant channels (Salt Pond E14, September 2010)

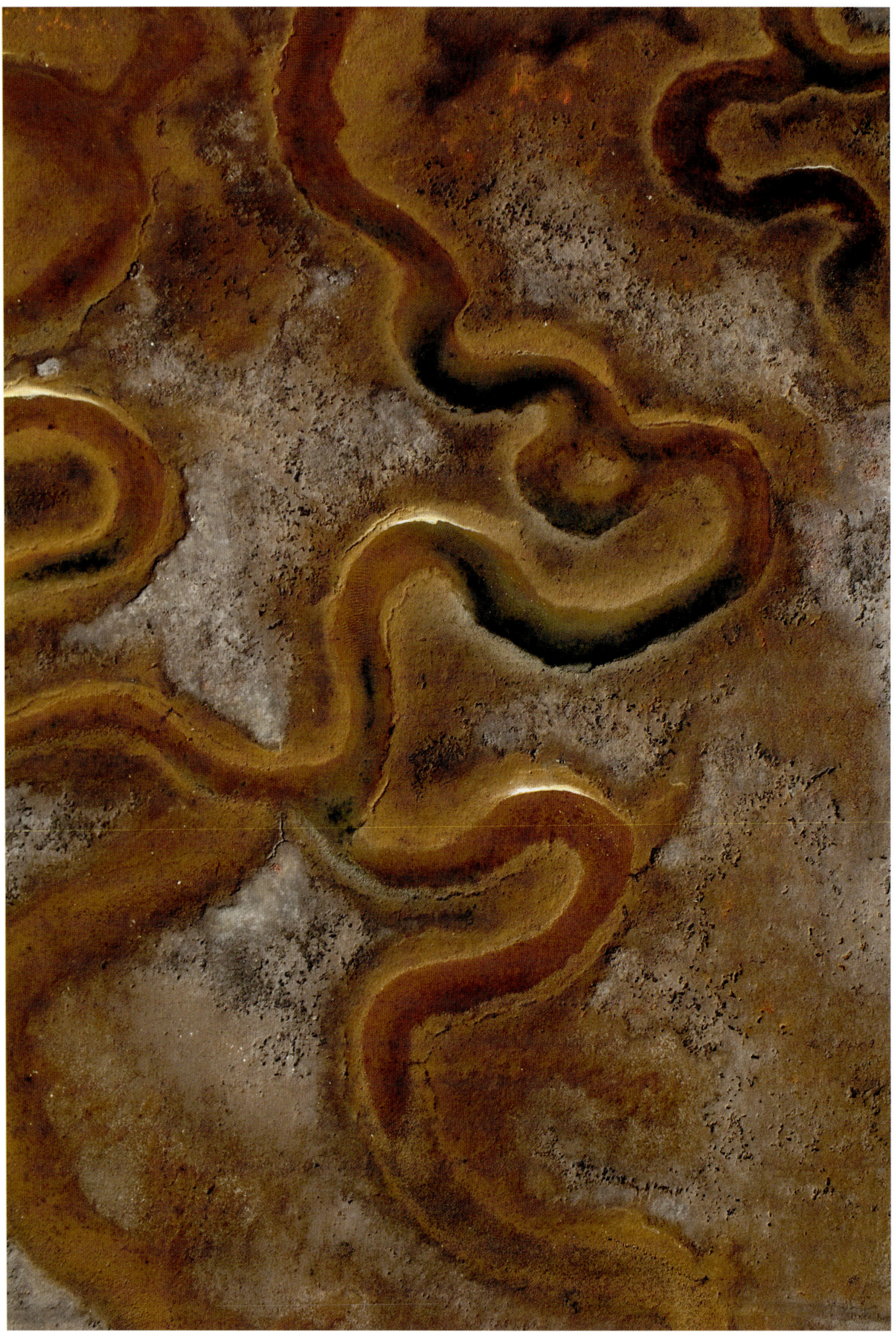

Plate 26. Bare channels (Salt Pond E8X, September 2009)

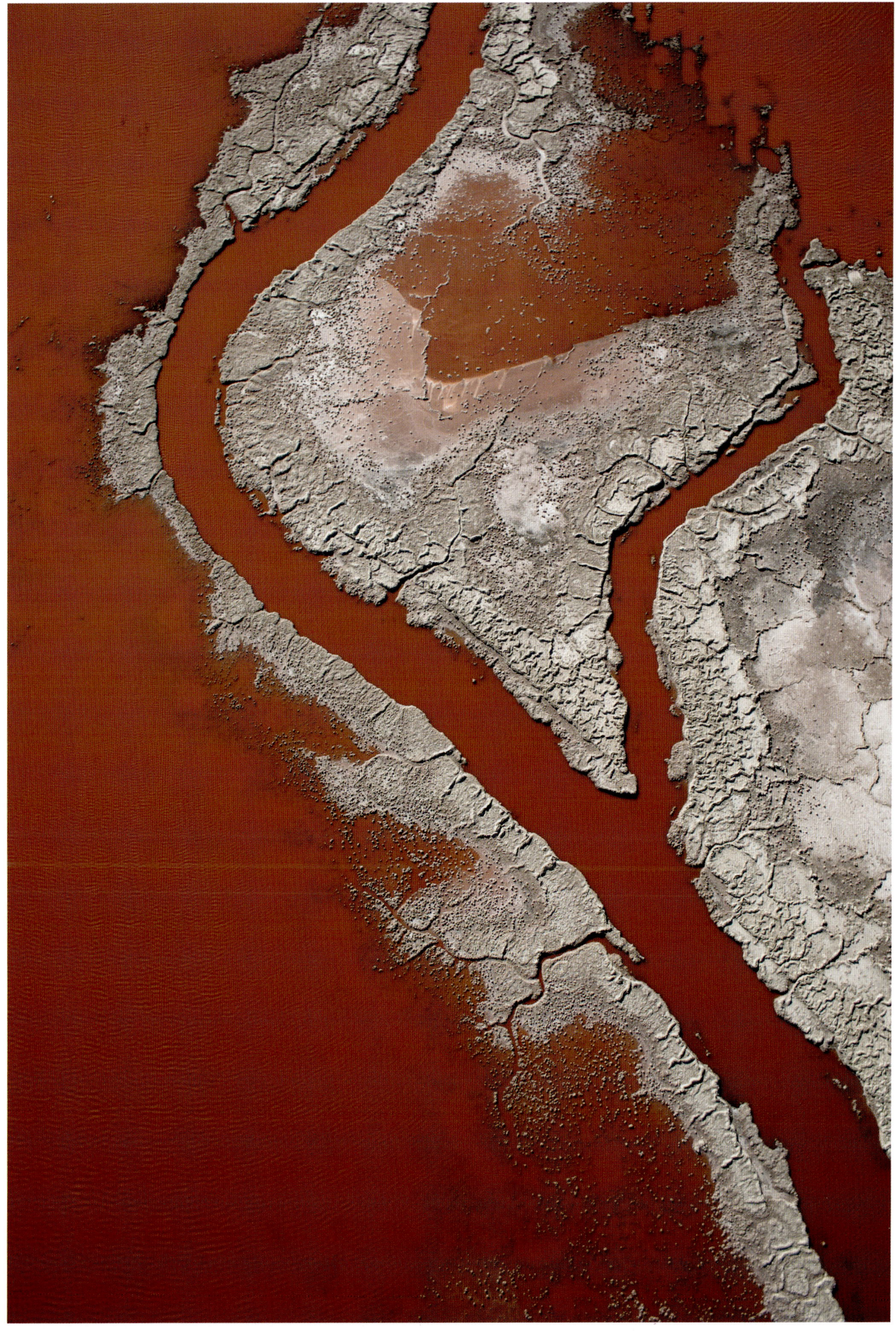

Plate 27. Bottom emerging as brine concentrates (Salt Pond M5, July 2008)

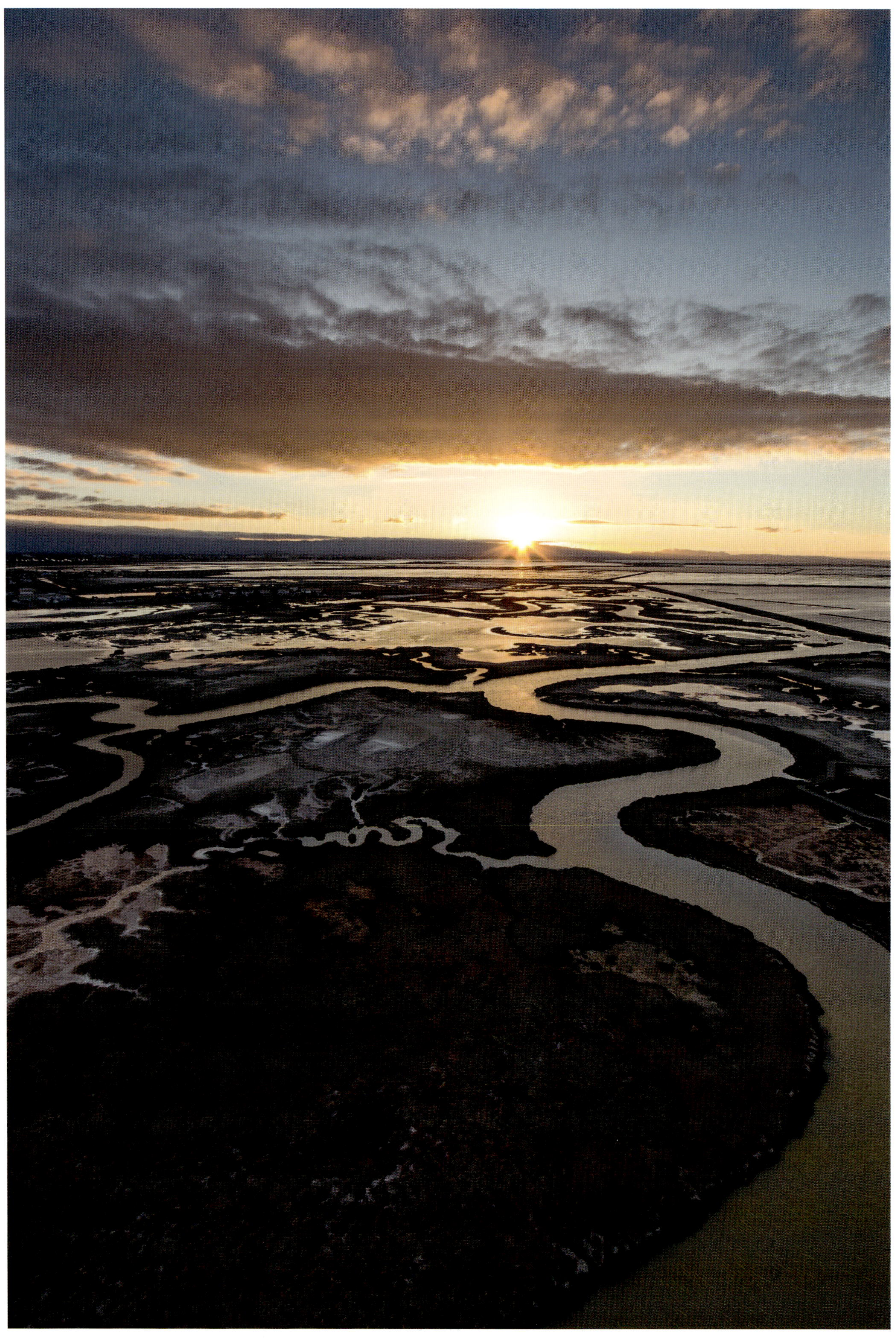

Plate 28. Another day ends (New Chicago Marsh, September 2011)

The South Bay's dormant marsh channels have a strange beauty. When viewed from above, their shape is immediately familiar, while the colors and textures are distinctly foreign. It is as though the channels have adopted a silly costume. The salt ponds are, by and large, devoid of vegetation. Old marsh channels appear in the ponds as brightly colored figures against the visual ground of the pond bottom. The ground is often abstracted by the buckled texture of gypsum, the calcium sulfate that precipitates in medium-salinity ponds. In other locations, the channels' features are softened by decades of sediment and erosion. As the oldest, baseline layer in the visible traces of the South Bay's past, it is ironic that the marsh channels are so visually dominant. But these are features that took centuries to establish and they seem to have a resilience borne of deeply belonging to the place.

I found myself searching out marsh channels, for as a photographic theme they proved every bit as entertaining as the landscape's colors. I quickly realized that the channels exist virtually everywhere bottom detail is visible in the South Bay, often hidden from ground view but quite evident from above. Impressed with the sheer scope of their distribution and wanting to know more about the original distribution of the marsh, I spent some time at UC Berkeley's Earth Sciences and Map Library reviewing old maps of the area. There to my delight I discovered the US Coast Survey maps from the 1850s— beautifully detailed maps of the South Bay marshlands made before significant human intervention.

The making of US Coast Survey maps, commonly called T-Sheets, is a story in itself. Fieldwork for the maps

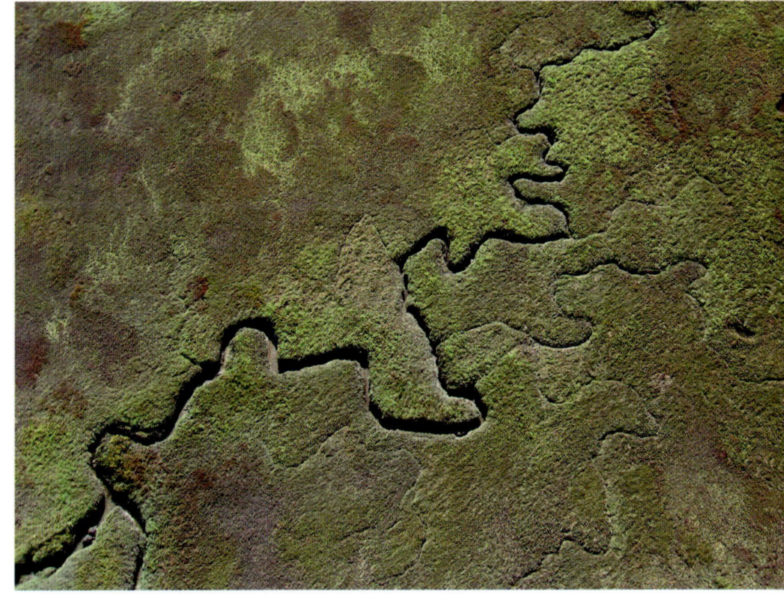

This section of undisturbed marsh at Dumbarton Point recalls the once vast stretches of open marsh along the East Bay shoreline.

was completed between 1854 and 1857; just eighty years after Anza first probed the South Bay shoreline and six years after Mexico ceded California to the United States. The principal surveyors, David Kerr and Augustus Rodgers (both in their mid-twenties) used relatively advanced plane table survey methods to map over one hundred square miles of South Bay tidal marshland. This they did

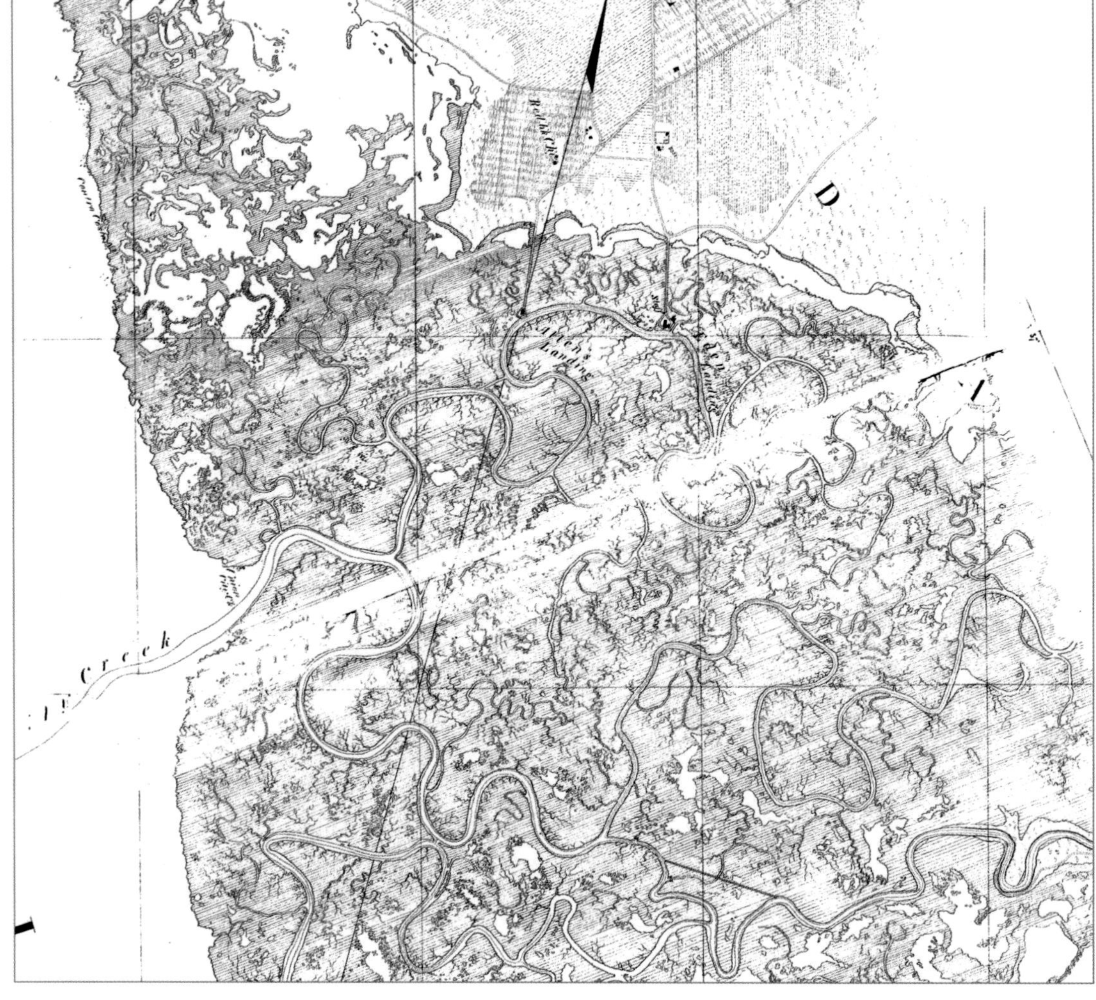

Detail from David Kerr's 1857 T-Sheet survey (t635) of the East Bay shoreline. Mount Eden Creek runs through the center of the map with Allen's Landing and Mt. Eden Landing plotted as evidence of early settlement at the marsh's edge. Courtesy of San Francisco Estuary Institute

while their contemporaries were abandoning all else to seek gold in the Sierra. The young surveyors produced twenty-seven sheets depicting the South Bay in remarkably accurate detail. The marsh channels plotted by Kerr and Rodgers align nicely with those visible in my aerial views. The Coast Survey T-Sheets convey a fine sense of how vast this marsh once was. And the knowledge that these channels still exist, at least in some form, seems somehow encouraging.

I have also captured images of marsh segments that survived undisturbed. Sections of live marsh exist along the bay shoreline near Eden Creek, on the Dumbarton Peninsula, and near Station Island. These, along with sections of exposed mudflat, serve as muted reminders of the landscape's former natural state. And then there are parcels that stand somewhere between the starkness of active salt ponds and subtleties of natural marsh. In fact exceptions to these extremes are both common and varied. The New Chicago Marsh near Alviso, a parcel that eluded development, is largely cut off from the tides—part marsh and part impoundment. The LaRiviere Marsh in Newark is well connected to the tides and features a returning mantle of marsh life where once a salt plant existed. Up north near Mount Eden Creek, adjacent parcels relate to the tides in a variety of ways, while at Ravenswood former salt ponds are held empty and fallow. These parcels, existing in various states between nature and industry, are becoming more numerous and represent purposeful investigations into alternative futures for our dormant marsh channels.

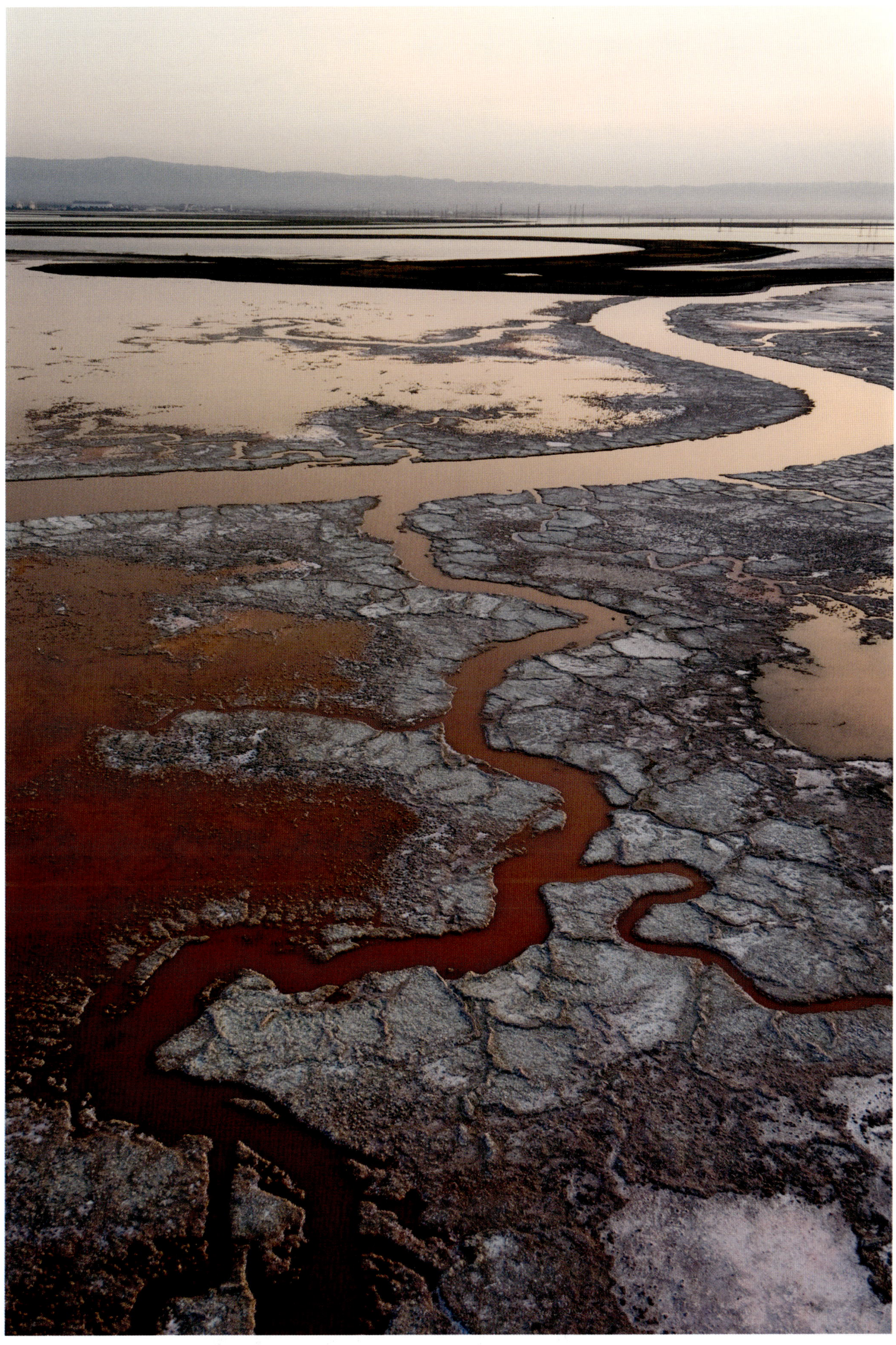

Plate 29. Red pond at sunset (Salt Pond M5 , July 2008)

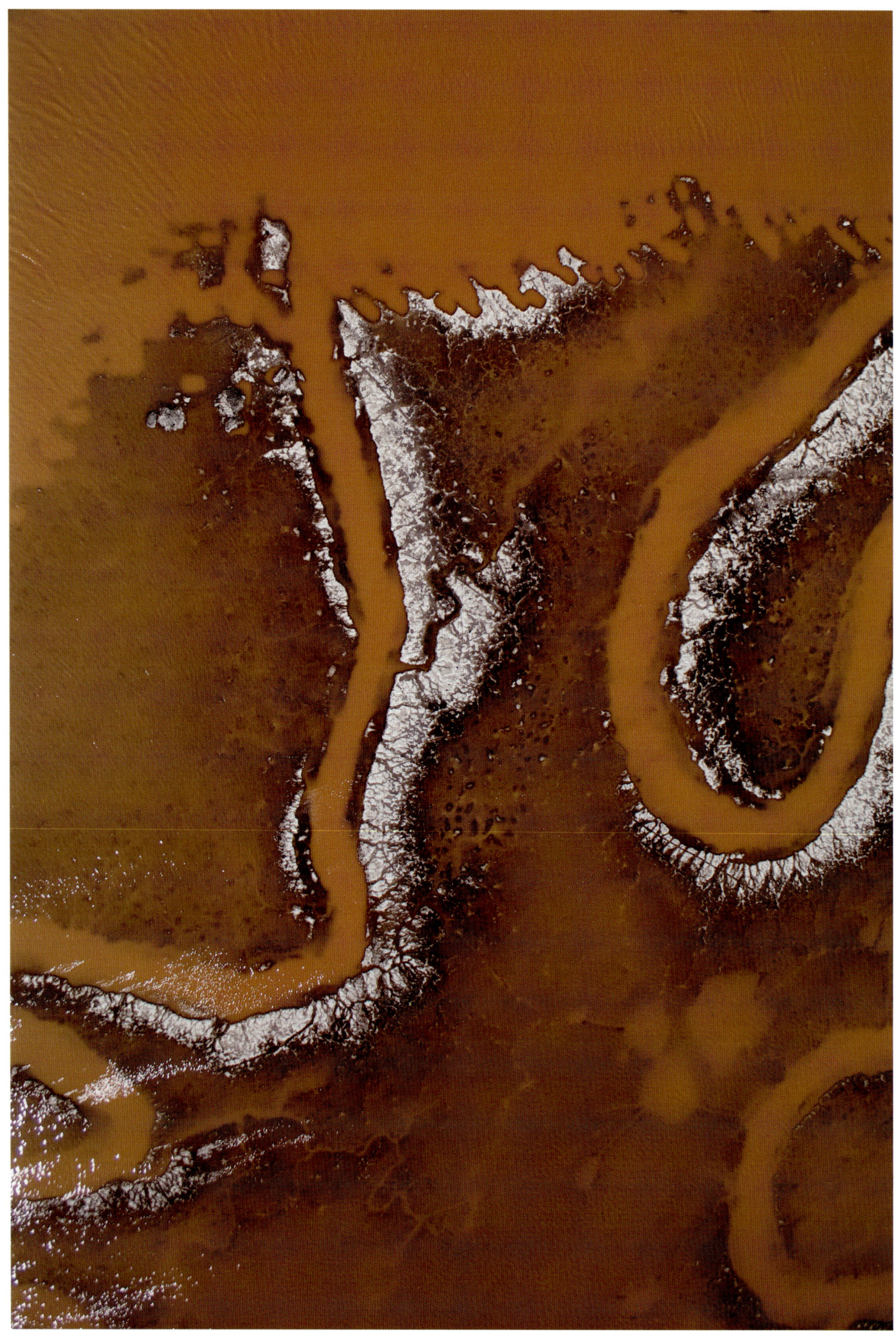

Plate 30. Channels and pond plateau (Salt Pond M4, July 2008)

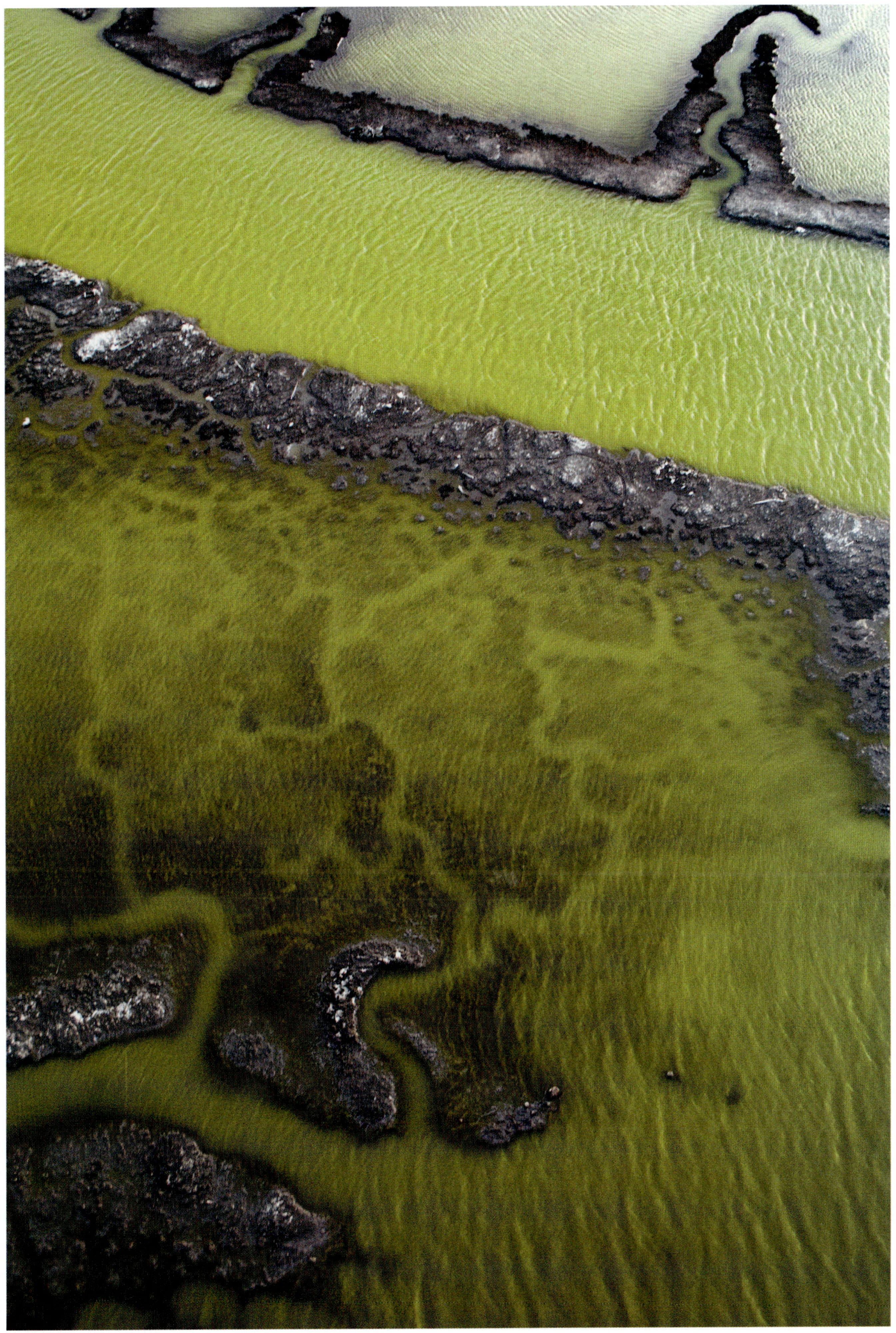

Plate 31. Raised edges (Salt Pond A12, March 2008)

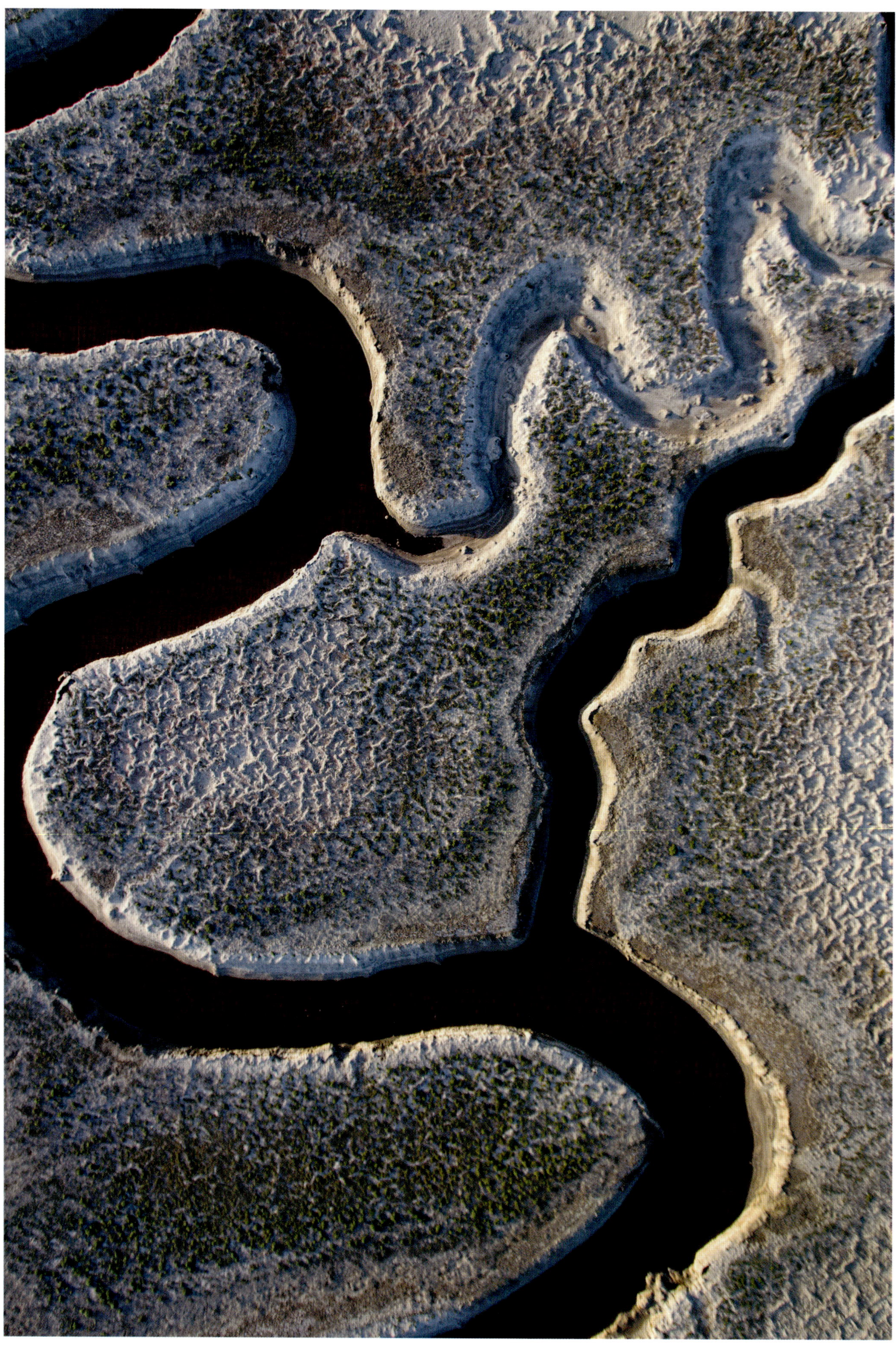

Plate 32. Distinct channels frozen in place (Salt Pond R5, May 2012)

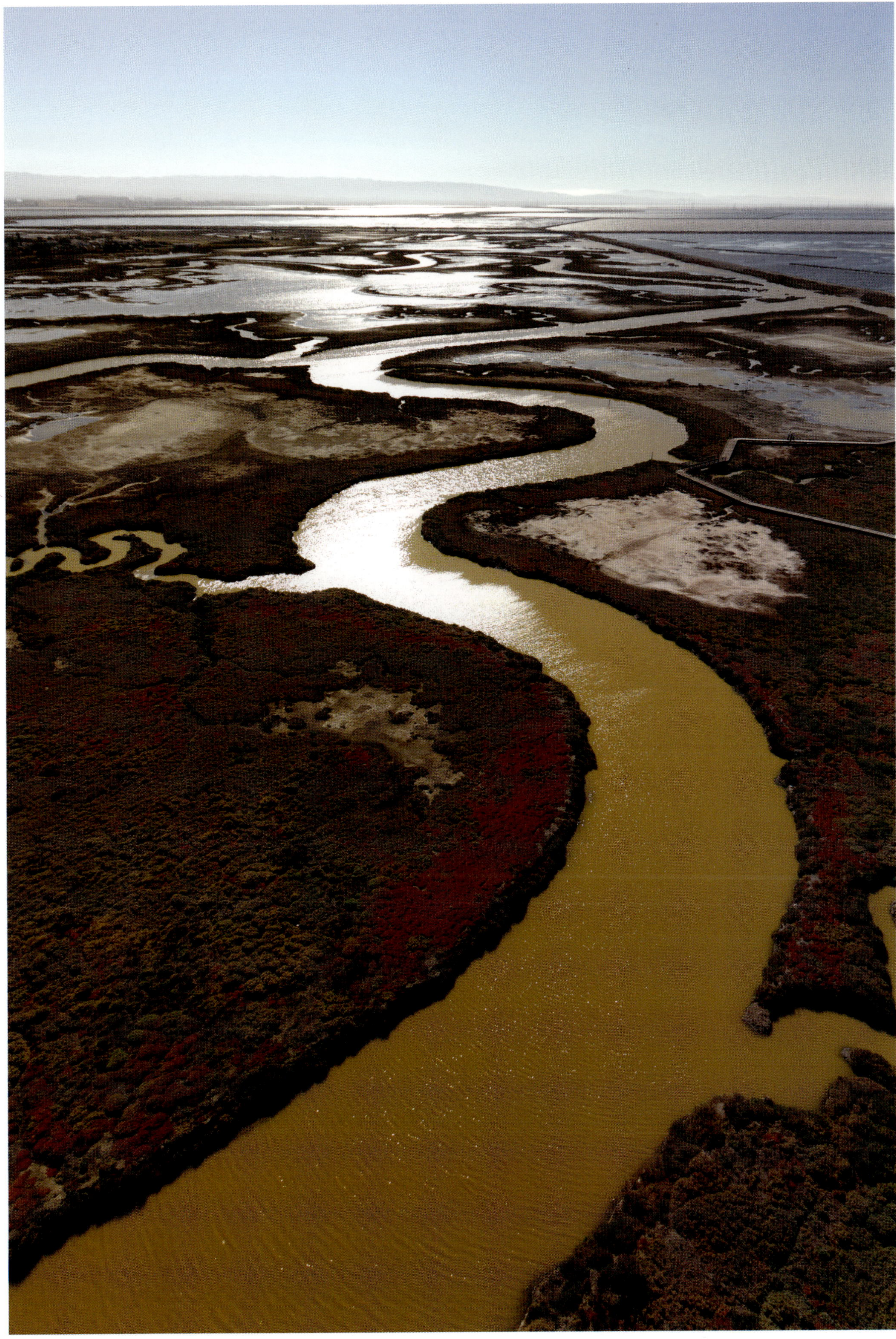

Plate 33. Still channel (New Chicago Marsh, August 2011)

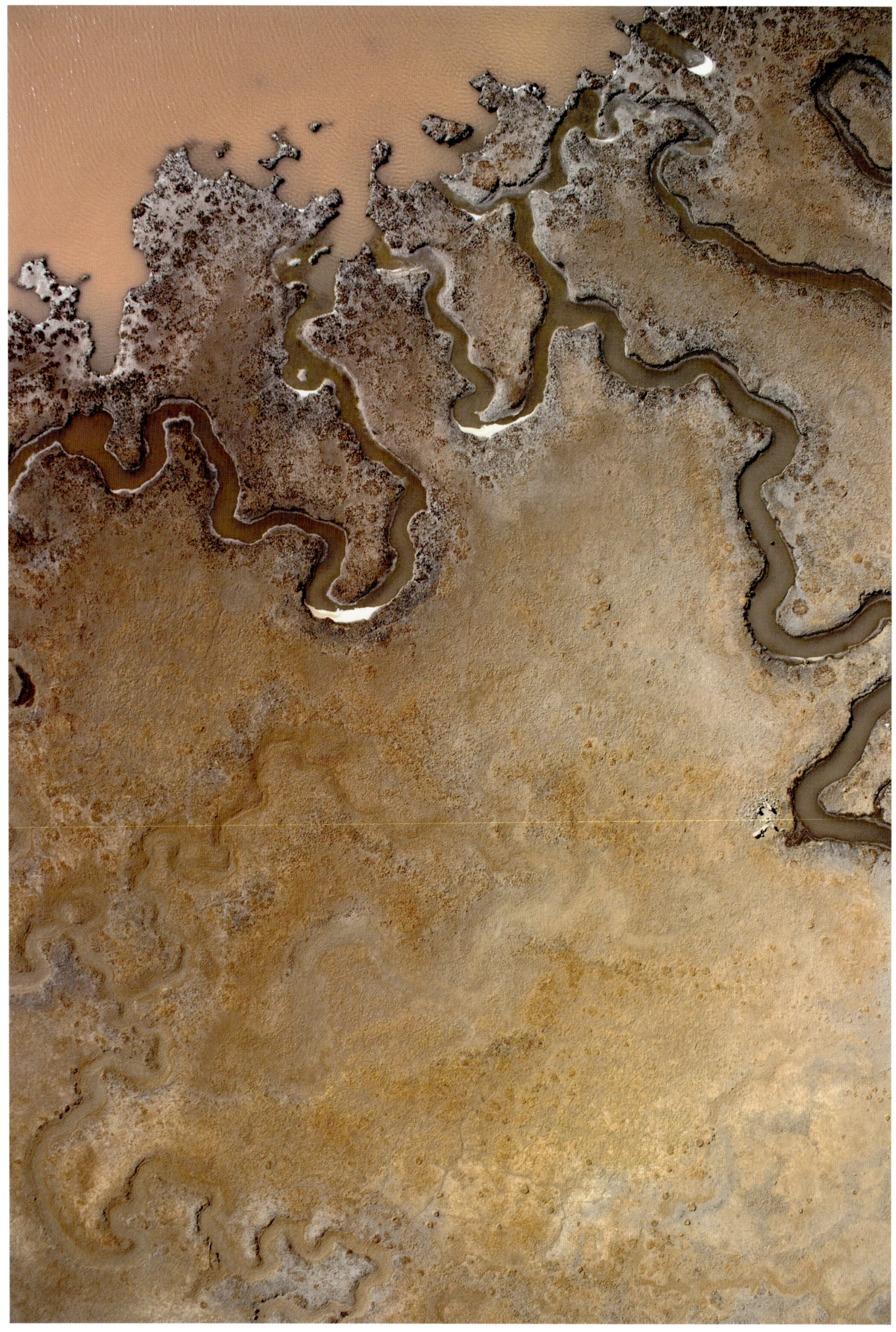

Plate 34. Familiar shapes, foreign texture (Salt Pond E8, October 2009)

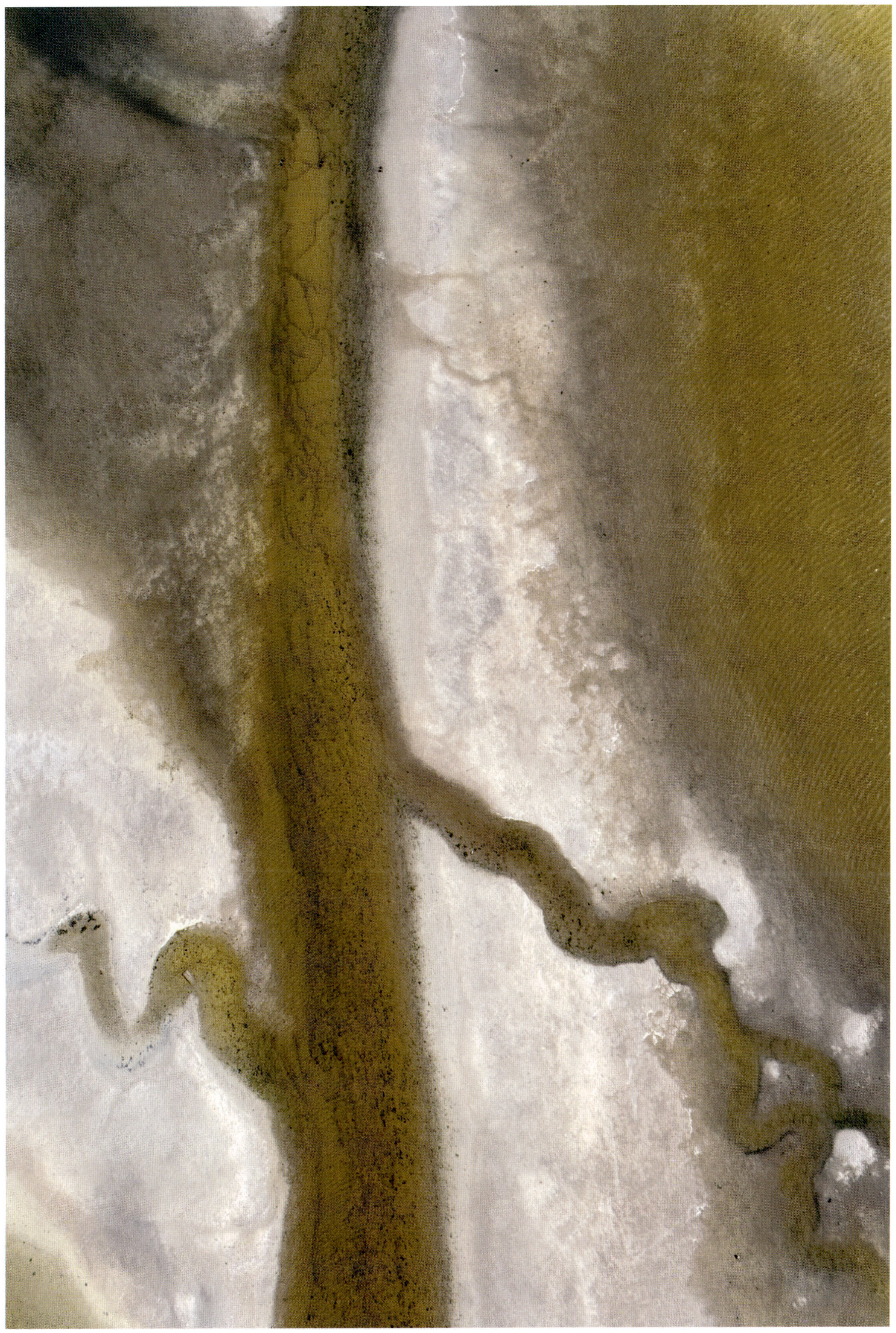

Plate 35. Channel revealed in drying pond (Salt Pond E4C, March 2010)

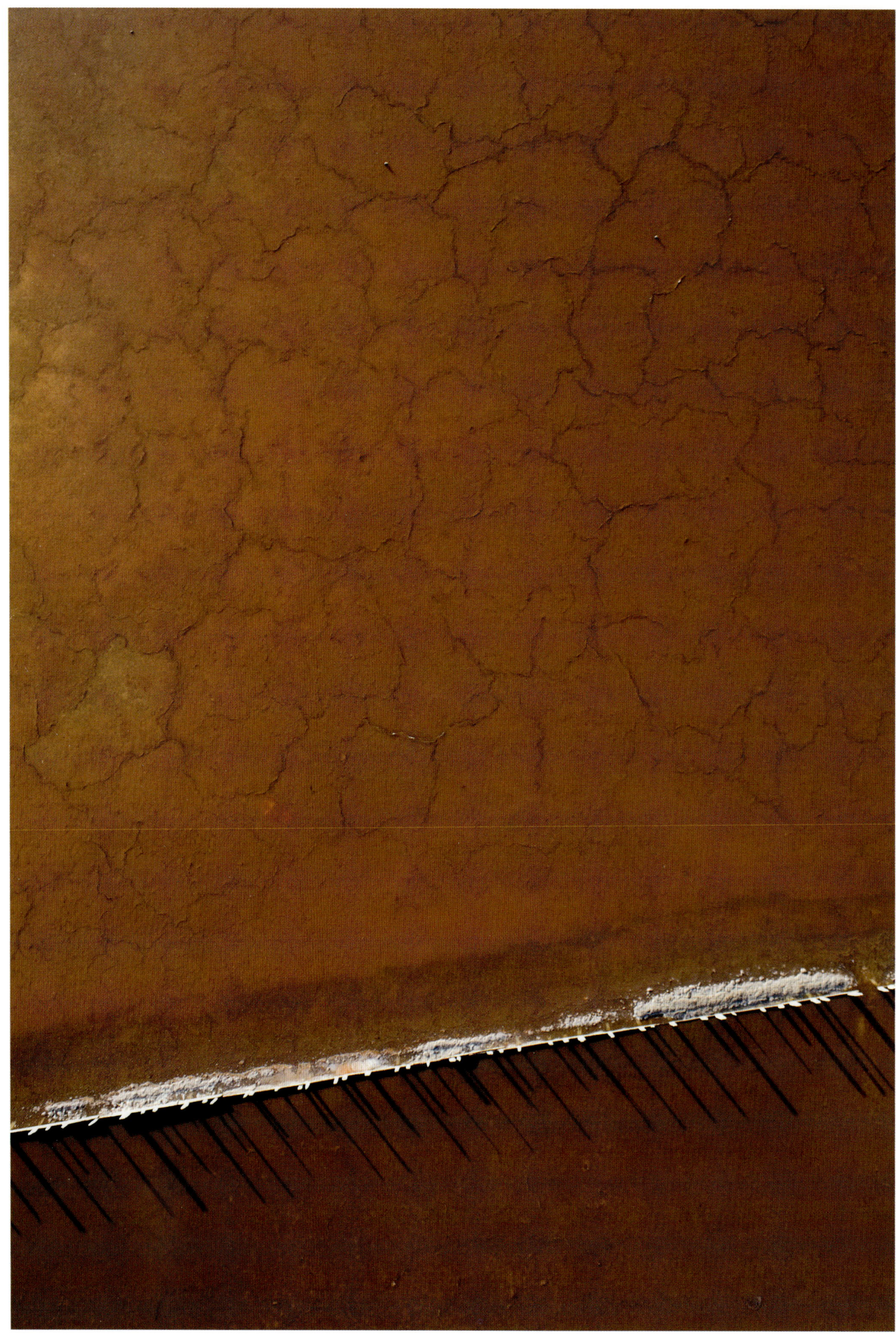

Plate 36. Former fence (Salt Pond E13, November 2008)

4: TRACES PAST

AT SOME POINT IN 1857 surveyor David Kerr must have stood on the newly established dock at Eden Landing—it was one of the triangulation points for his T-Sheet map of the East Bay wetlands. From this vantage point, gazing south, he would have seen a continuous stretch of marsh averaging three miles wide and running to, and beyond, the Coyote Hills five miles distant. It would have been hard for Kerr to imagine that this vast, uninterrupted network of sloughs, channels, and pannes would essentially disappear over the next few decades to be replaced by a patchwork of saltworks resembling small farms. It would have seemed stranger still that the small saltworks resulting from this unimaginable transformation would themselves disappear after a few more decades and that further transformations would follow at similar intervals. The wetland map Kerr was completing showed but a few human settlements on the bay's edge, including the first boat landings—Thompson's, Allen's, Johnson's, and Eden—built to connect East Bay farming communities to the San Francisco market.

A current-day visit to the disused salt ponds along Mount Eden Creek yields, from the ground, an impression of barren flat fields embedded with a seemingly random scattering of ruins and artifacts. Yet my first aerial photographs of the area revealed a subtle, complex order to the rich quilt of visible traces and patterns. The scenes juxtapose faint, meandering traces of natural features dating back to Kerr's view of the marsh with a chaotic overlay of constructed features that, in aggregate, are complex enough to defy immediate comprehension. Over time I developed an awareness that these traces superimposed patterns of the original marsh, interventions of successive stages in the salt industry, the gravitational pull of different transportation systems, and elements tied to collateral uses of the wetlands. What an interesting visual puzzle! The levees and other features from the early nineteenth-century saltworks provide the finest grain of subdivision, patterns generally more faint than other features. The older saltworks were located adjacent to systems of transportation, first nautical and later rail, and organized around natural features that challenged and constrained growth.

Patterns in the landscape also reflected several social epochs. Traces in my images belong to periods underpinned successively by the bonds of local communities, by freewheeling entrepreneurial capitalists, by relatively distant and mechanized industry, and by environmental stewardship at a societal scale. My attempts to understand the landscape's visual traces inevitably led to curiosity about everyday life at different stages in the development of this place and fired the imagination about what it might be in the future. That these traces exist at all is great testimony to the resilience of the landscape and evidence of its potential to accommodate sweeping change.

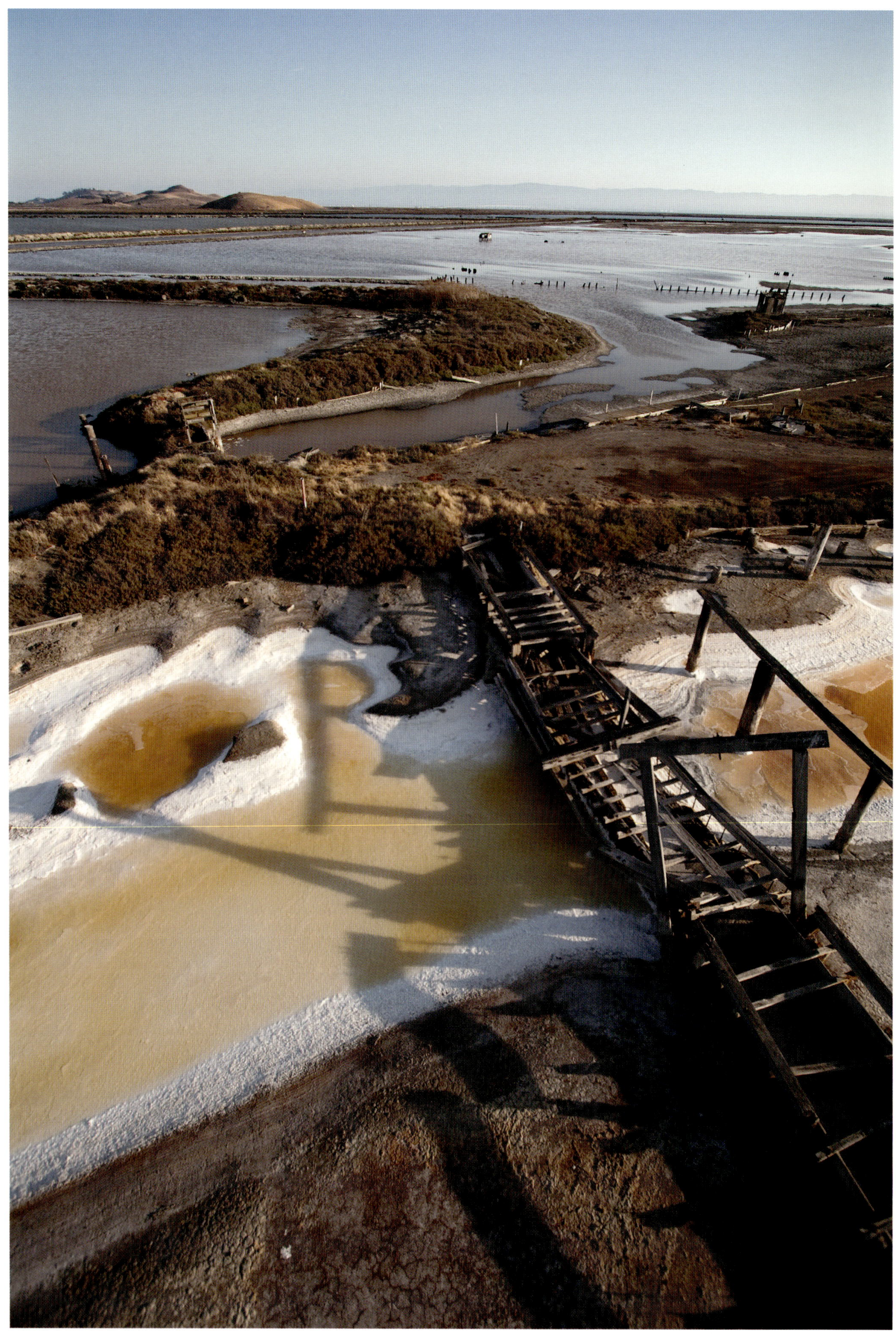

Plate 37. Collapsing bridge (Former course of Alameda Creek, August 2009)

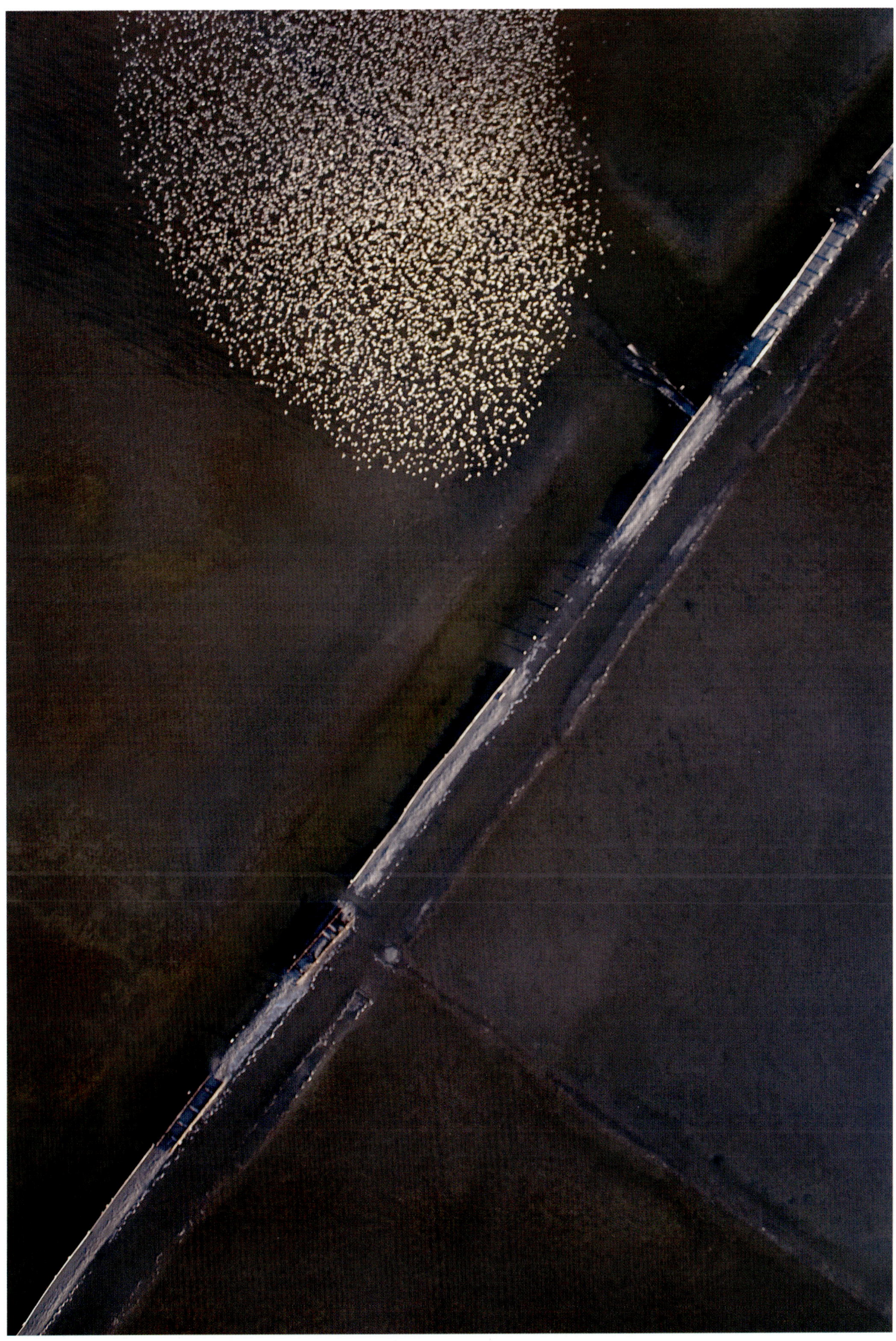

Plate 38. Bird swarm at sunset (Salt Pond E13, November 2009)

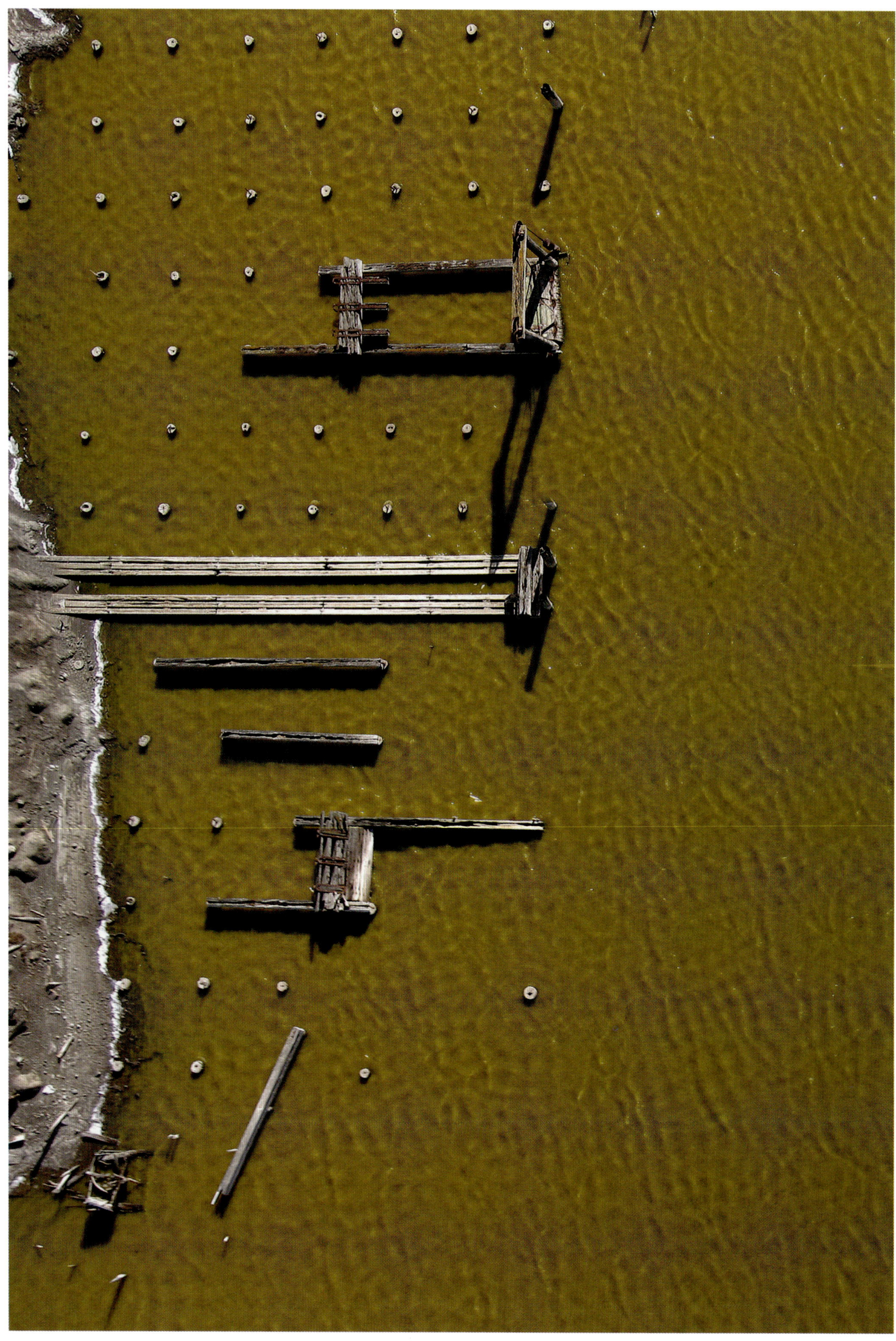

Plate 39. Landing ruin at Dumbarton Point (Salt Pond N3, October 2005)

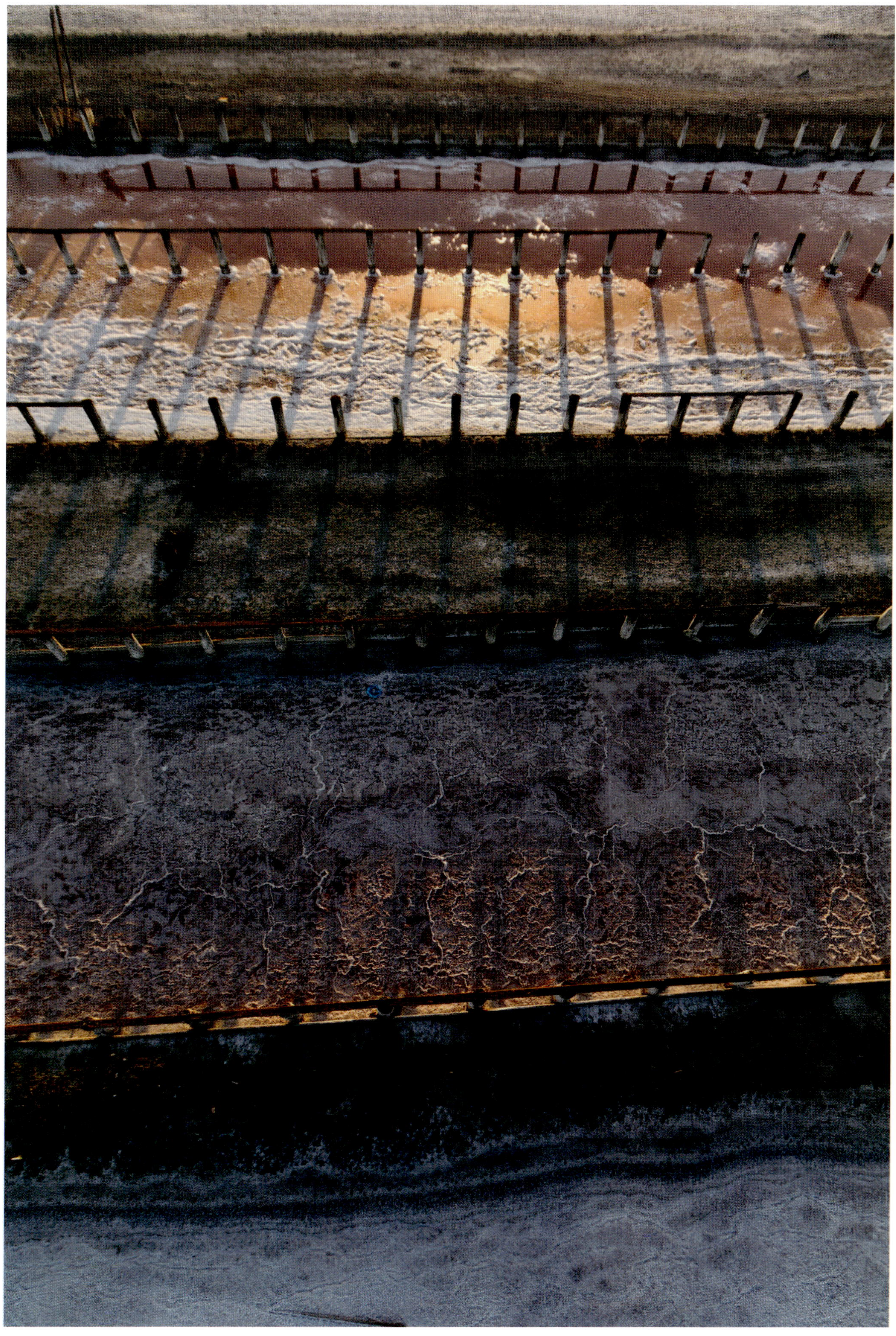

Plate 40. Pile array at sunset (Oliver Salt Works ruins, September 2010)

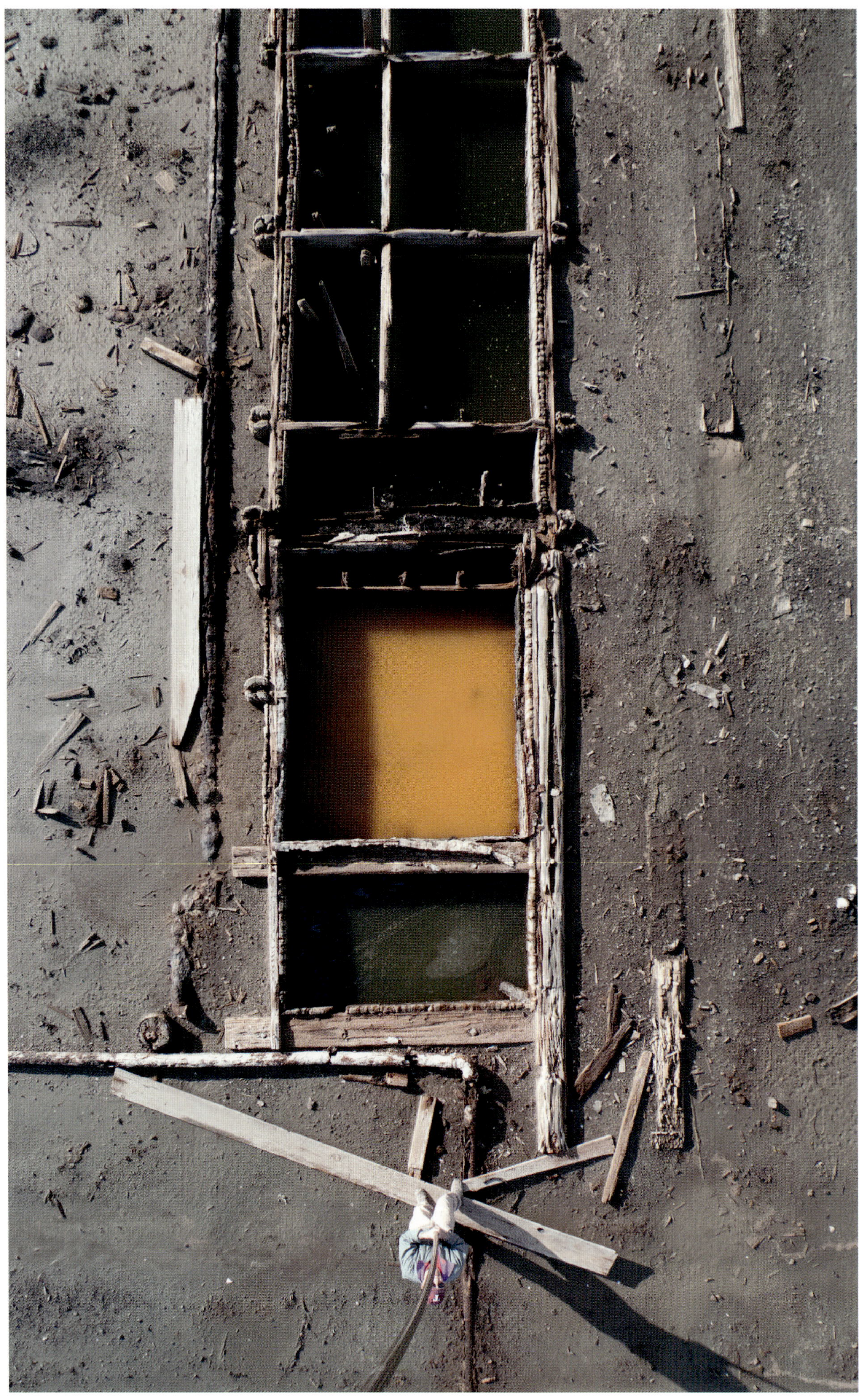

Plate 41. Sub-grade holding tanks (Oliver Salt Works ruins, January 2009)

Deciphering the patterns in my aerial photographs of Eden Landing involved iterative cycles of wandering the landscape then wandering the library. My understanding of the salt industry benefited greatly from John Sandoval's *Mt. Eden: Cradle of the Salt Industry in California* and the marvelous *Salt in California*, William E. Ver Planck's 1957 profile of salt making in California and its history. Salt production in the Bay Area evolved through several distinct stages, each with its own pattern and scale.

The Bay Area climate is particularly well suited to salt making. Reliable westerly breezes and rainless summers combine to allow shallow ponds to lose more water by evaporation than they gain from precipitation. This net evaporative loss occurred naturally along the East Bay shoreline, where the highest tides of the year would strand salty bay water in shallow pools. Subsequent evaporation led to naturally formed crystallized salt, with modest amounts being harvested before the arrival of Europeans to serve local consumption and trade.

Salt production was taken to the next level by actively constructing small ponds in the wetlands. The process involved tedious hand labor, digging impervious marsh mud from a "borrow ditch" to create adjacent levees. Levees were arranged to deny tidal flow to a series of enclosed basins, which were then actively managed to increase salinity and ultimately precipitate salt. The first commercial venture of this sort was attributed to John Johnson, a Scandinavian immigrant who settled near the community of Mount Eden and created the first mud levees in 1853. Johnson's nascent saltworks made a modest profit, largely selling salt for tanning hides and processing leather. His achievement occurred at a time when the federal and state governments were promoting the

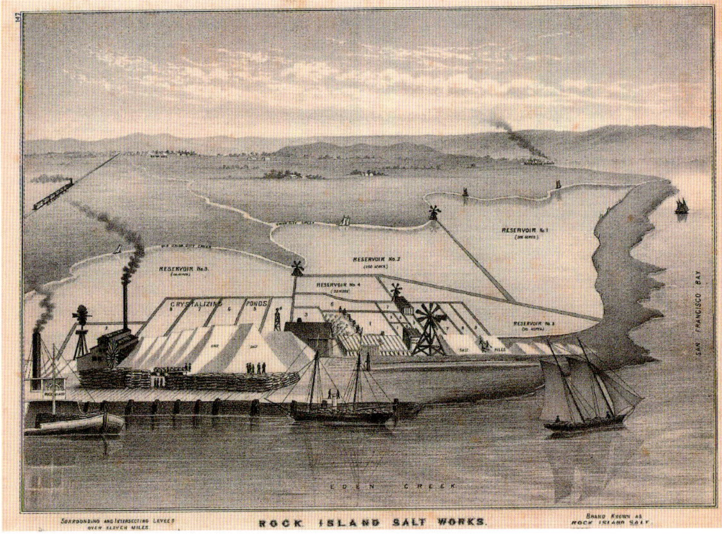

The Rock Island Salt Works: this depiction, while surely aggrandized, conveys a sense of the general layout and constituent parts of the smaller nineteenth-century saltworks. The view also features the Coyote Hills, the newly created railroad, and smoke rising from Dumbarton Point. From *Official and Historical Atlas Map of Alameda County, California* (Thompson & West, 1878), courtesy of the David Rumsey Historical Map Collection (www.davidrumsey.com)

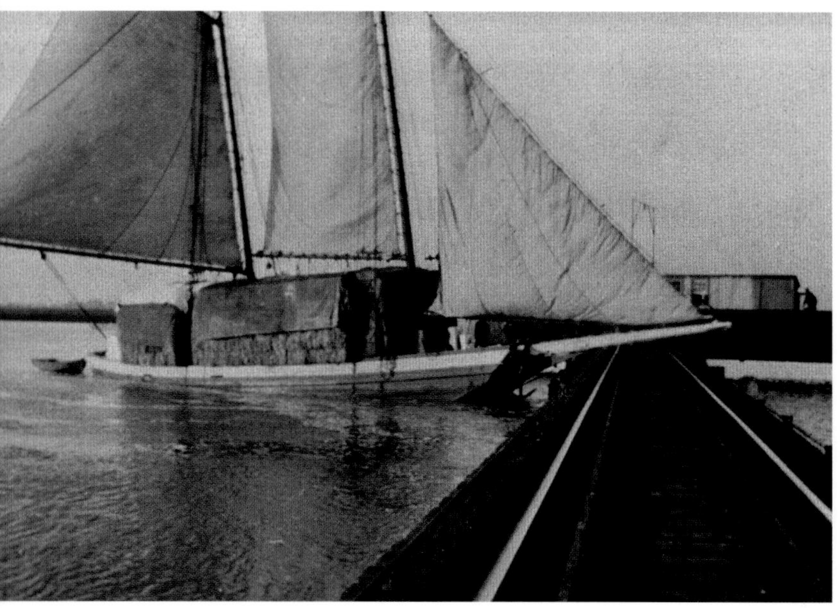

A hay-laden scow schooner nudging the rail bridge. In the nineteenth century, over 250 scow schooners served as utility delivery vessels in San Francisco Bay. Favored for their simple design, shallow draft, and flat bottoms that were easy to beach, the vessels became less common as freight traffic shifted to the rail lines. Courtesy of Don Edwards San Francisco Bay National Wildlife Refuge

sale of parcels of up to 320 acres of "swamp and overflowed land" to individuals willing to improve the land. San Francisco's gold rush population explosion and the use of salt in metallurgy led to increased demand for salt, which led in turn to more settlers building saltworks. Within three decades, over two dozen saltworks were constructed in the marshlands near Mount Eden.

The saltworks established between 1850 and 1900 were generally modest in scale, similar perhaps to family farms. They were established on land grants of a few hundred acres or less and revolved around the capacities of human labor. Their levees, small and economical due to the means of construction, were barely up to the task and often failed. Evaporator pond layout was constrained by natural features: the simple levees lacked a capacity to span even moderate marsh channels, and the natural forces of tide and wind were the only means to move brine from pond to pond. Brine flow on the sites eventually led to compact crystallizer beds, which were generally adjacent to a work's residences, processing sheds, and boat landing. Descriptions of John Johnson's prototype saltworks include a wharf and warehouse on pilings with a house nearby, a modest estate later expanded to include a hotel and picnic ground for hunters. There is an impression of the saltworks existing like tiny city-states: often cooperative, sometimes competitive, most certainly sharing social bonds within several layers of community.

John Johnson's provision for a boat landing was common to saltworks of the time. In the nineteenth century, the Bay Area was a decidedly nautical place. Sailors would skillfully guide two-masted, shallow-draft scow schooners up the winding East Bay marsh channels to boat landings strategically situated to link dry land with navigable waters. The open marsh of the day provided a large tidal prism, water flow that kept the major marsh channels scoured and navigable. Up and down the East Bay, boat landings were situated on the navigable limits of every sizable slough, their locations and names still evident on contemporary maps. The landings handled passengers, agricultural produce, redwood logs, animal products, and salt. As a product, salt was heavy enough to warrant close proximity between production and shipping; thus many saltworks were contiguous with boat landings.

As time passed, the early saltworks gained sophistication. Salt was at first milled in San Francisco, but the East Bay works established local facilities. With great ingenuity, salt makers developed a variety of wind-driven pumps, the most interesting being Andrew Oliver's 1870 Archimedes screw model. In a stiff summer breeze this pump could raise fifteen hundred gallons of brine a minute, a valuable feat that made the pumps a common fixture across the South Bay.

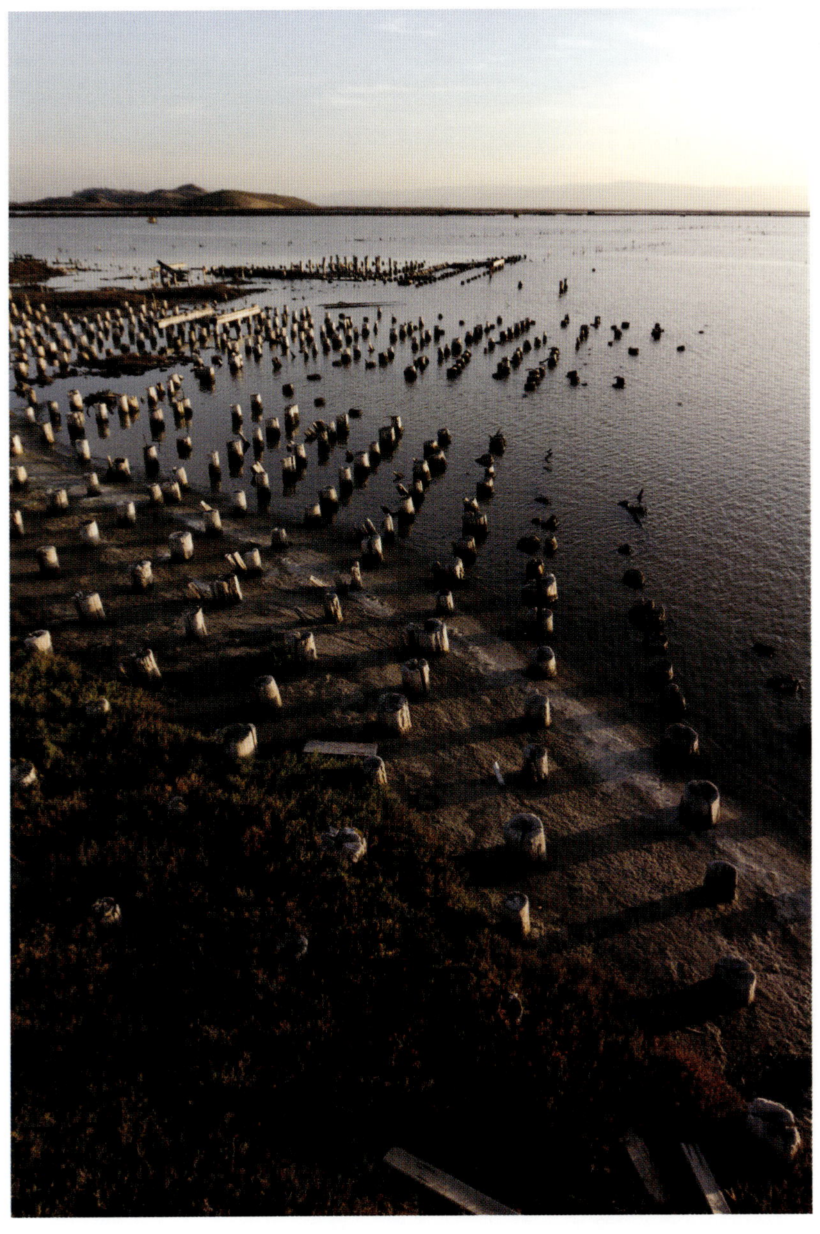

This array of piles once supported a wharf in the marshes just east of Alvarado, the site of the Union City Saltworks, located at the navigable limit of the old course of Alameda Creek. The creek has since been relocated to a flood control channel.

As the nineteenth century came to an end, there was a trend toward larger salt companies as the more successful works acquired their neighbors' property in deals often founded on community relationships. Still, a reliance on human labor and simple technologies limited any given saltworks to a modest scale. In aggregate they now covered perhaps half of the East Bay marsh, a constellation of small enterprises constrained by natural features. This would soon change.

Traces of these early works are quite recognizable in my contemporary aerial views. Hand-built levees from a century ago, only a few feet high when new, have weathered down to but a few inches. Yet these levees and their borrow ditches are still quite visible, providing interesting evidence of how and where the early saltworks were arranged. They are also testimony to the diminutive scale of the original works and the manner in which they conformed to constraining marsh channels. The photographs capture other features of the early saltworks—low fence lines, remnants of raised troughs for brine transport, the skeletons of old wind pumps, and the small straight-edged crystallizer beds. Here and there are more elaborate ruins of former saltworks, places where little remains above knee height but the layout of the processing facility is clearly evident in foundations and earthwork. The old Union City Salt Works near Alvarado and the Oliver Salt Works on Mount Eden Creek are standout examples. The latter was operated into the 1920s. Elsewhere previous works have faded to more subtle hints.

The nineteenth-century boat landings left traces that are often more difficult to visualize. The Union City Salt Works landing survives as a large ruin of pilings situated on a branch of Alameda Creek that ceased to exist many decades ago when a straight-line flood control channel supplanted natural drainage. Jarvis Landing on Newark Slough began operation as the embarcadero for Mission San Jose and has offered continuous access to the slough since the early 1800s. You can launch a kayak at Jarvis Landing today next to the few piles remaining from its heyday as a port. Piles are all that remains of landings on Mount Eden Creek, sitting incongruously along a once navigable channel now filled with silt. The most preserved of the landings is visible right at the eastern anchorage of the Dumbarton Bridge in a salt pond that was once the course of Beard's Creek. This landing appears on maps as late as the 1930s.

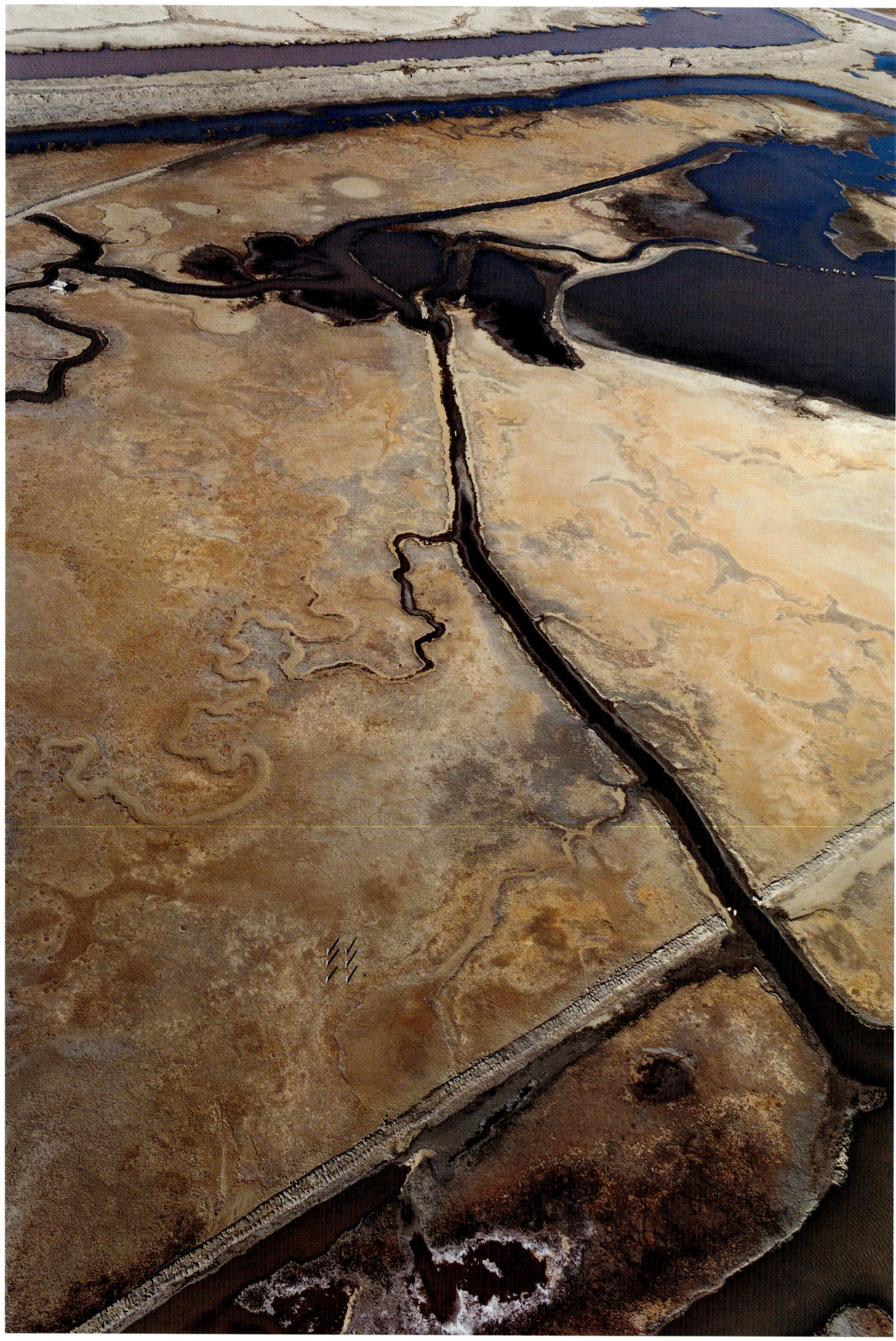

Plate 42. Faint traces of early works (Salt Pond E8, October 2009)

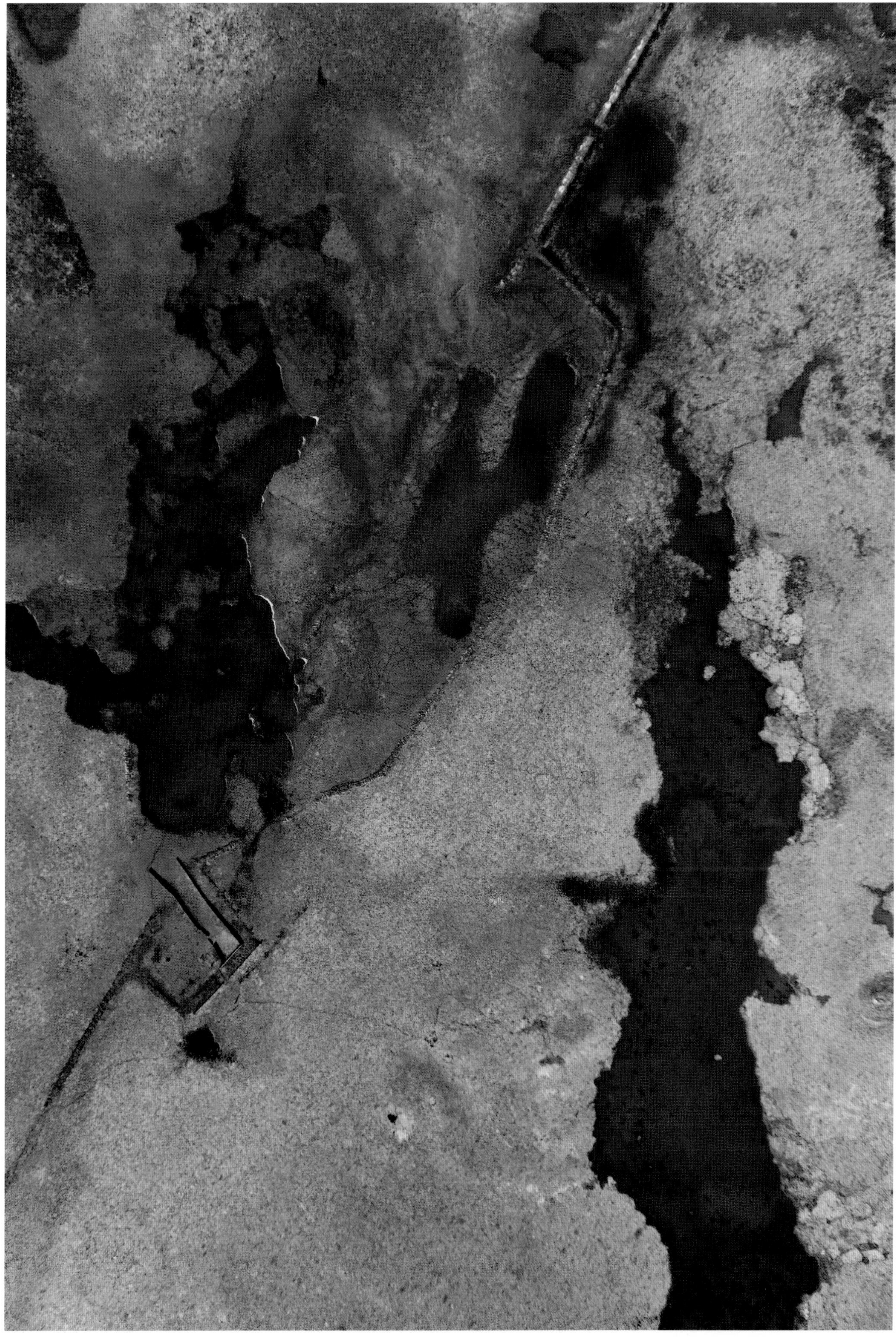

Plate 43. Foundation outlines in cyanobacteria mat (Salt Pond E8X, August 2011)

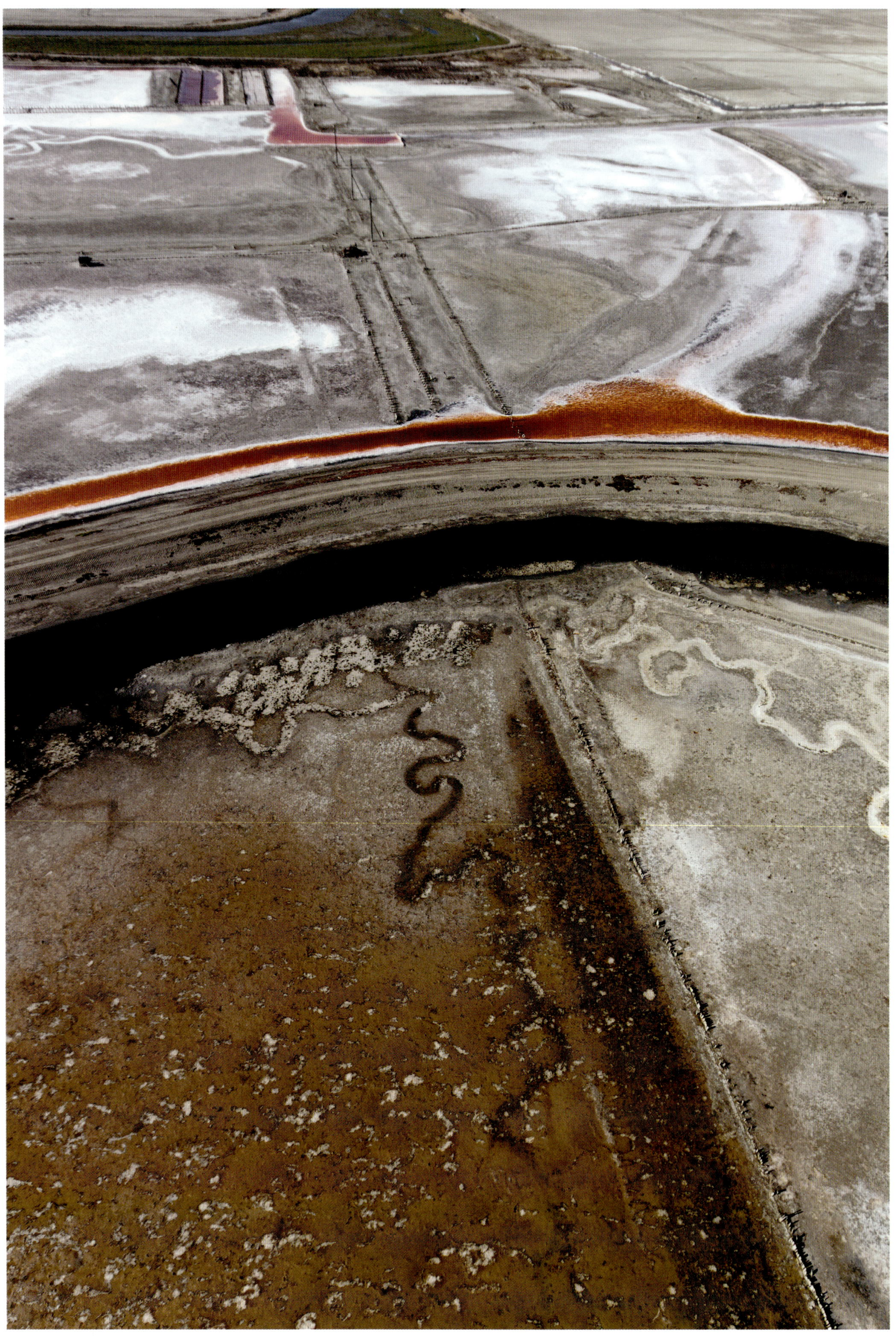

Plate 44. Evidence of early brine distribution network (Salt Ponds E14 and E13, August 2011)

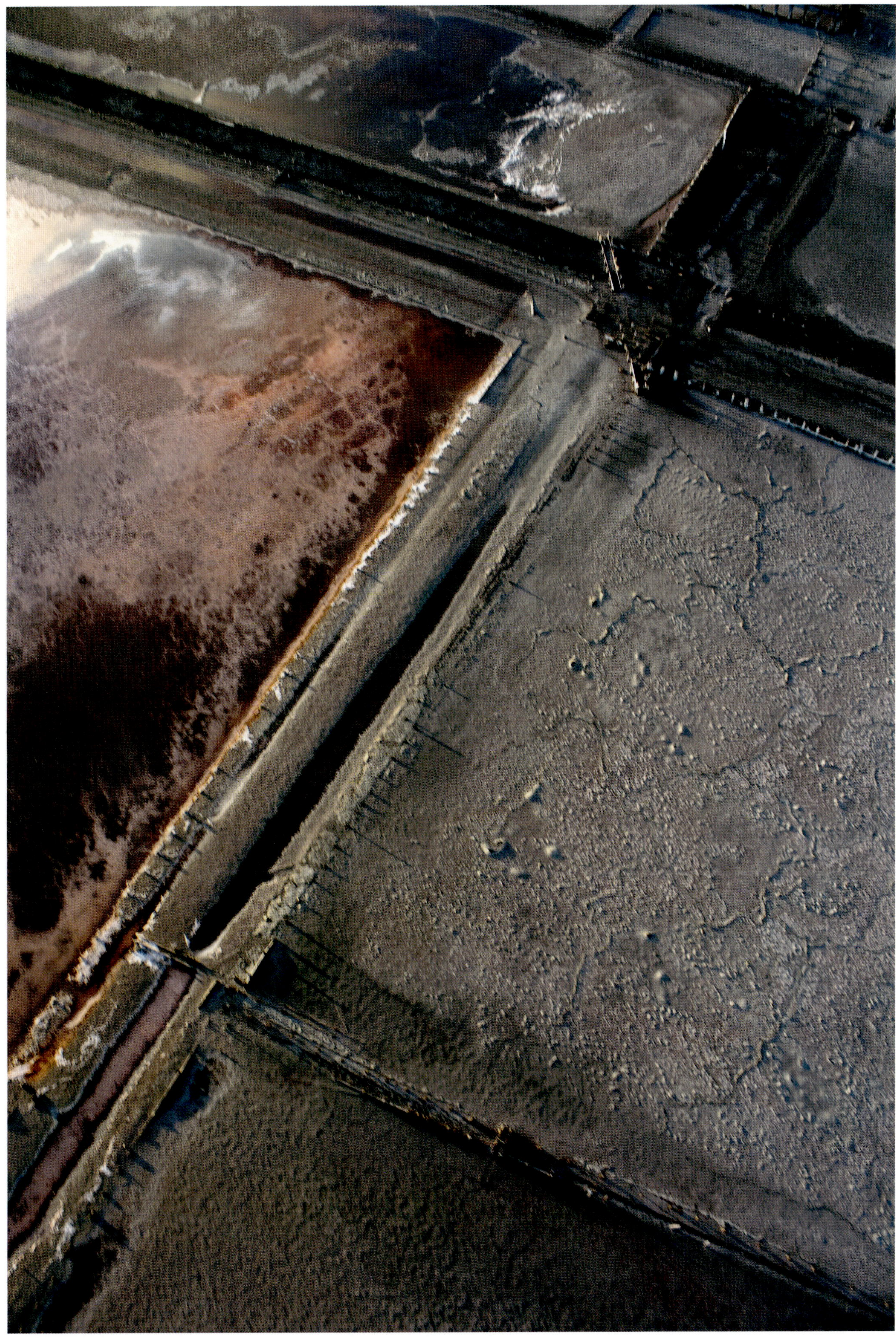

Plate 45. Former crystallizer bed network (Salt Pond E13, September 2009)

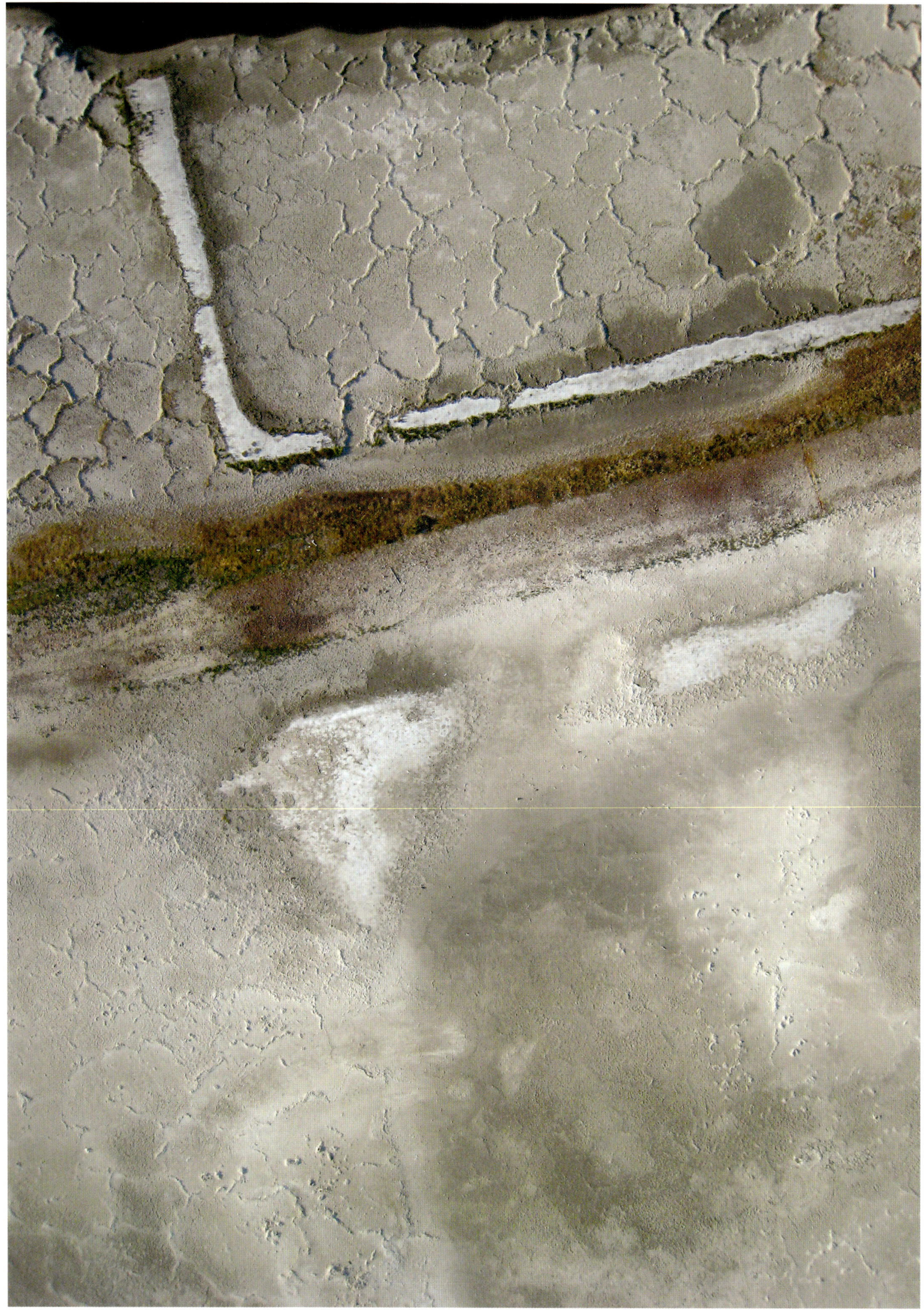

Plate 46. Complex traces in pond bottom (Salt Pond A22, November 2008)

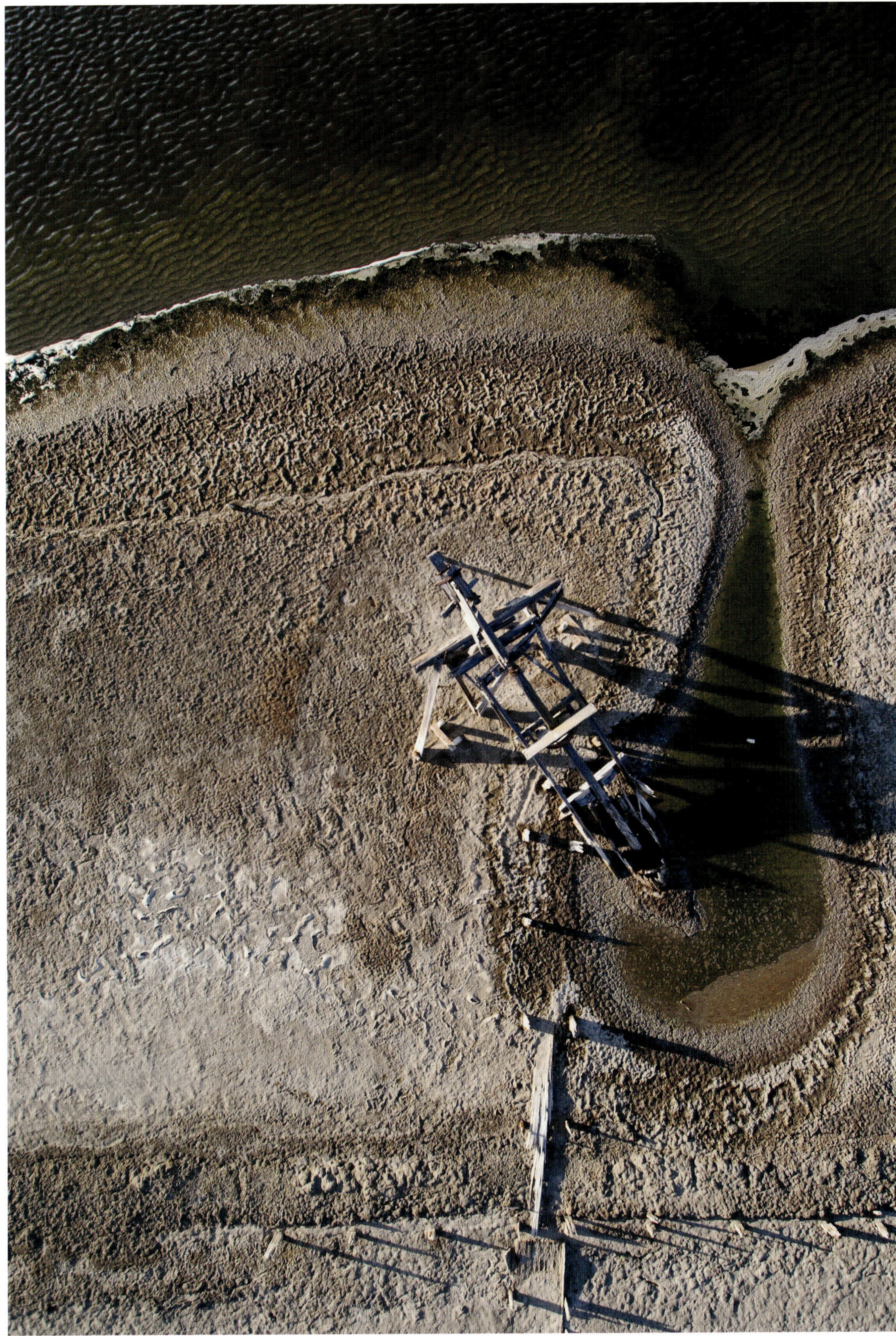

Plate 47. Archimedes wind pump ruin (Salt Pond E14, September 2008)

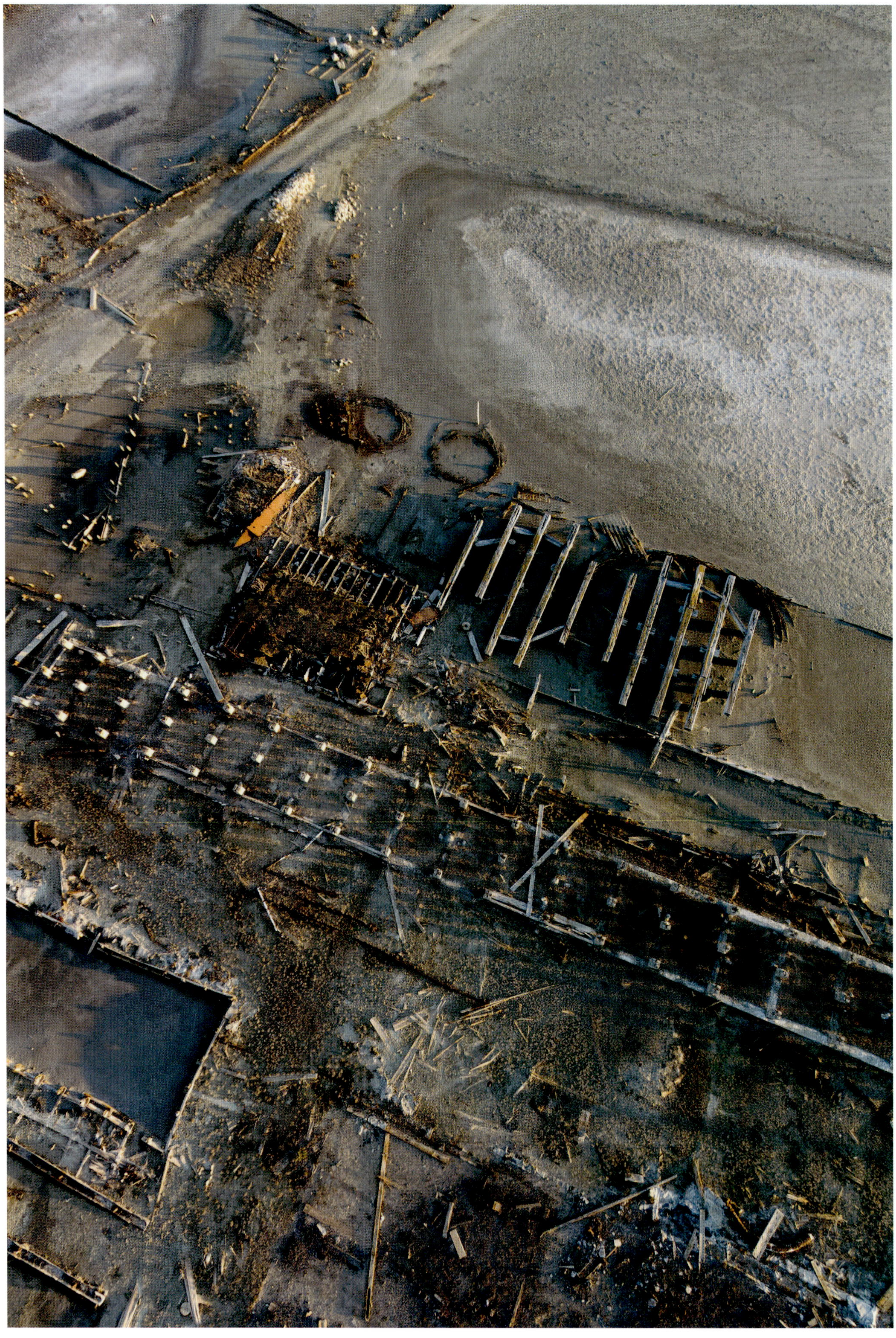

Plate 48. Foundations for large holding tanks (Oliver Salt Works ruins, October 2011)

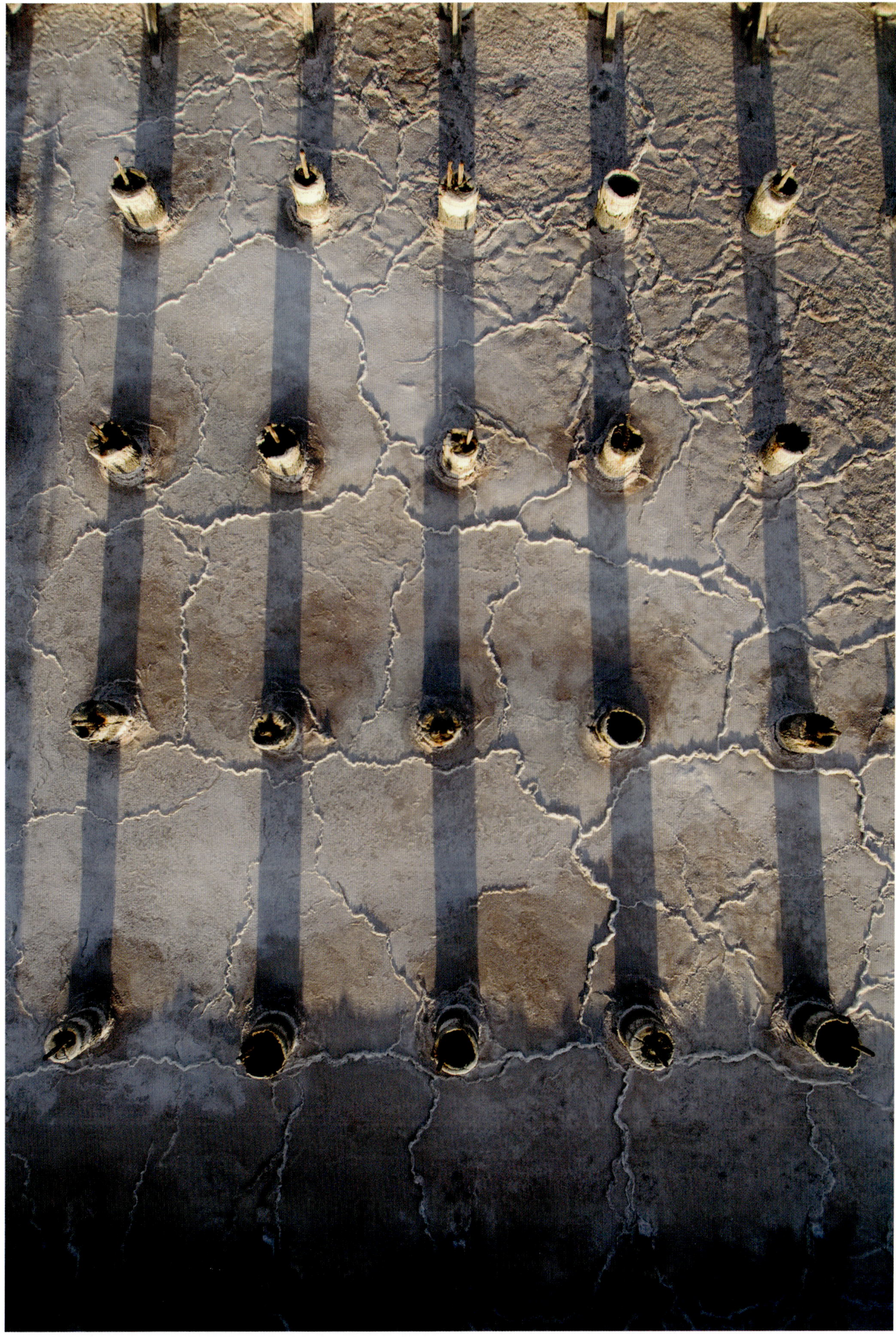

Plate 49. Piling array (Oliver Salt Works ruins, September 2008)

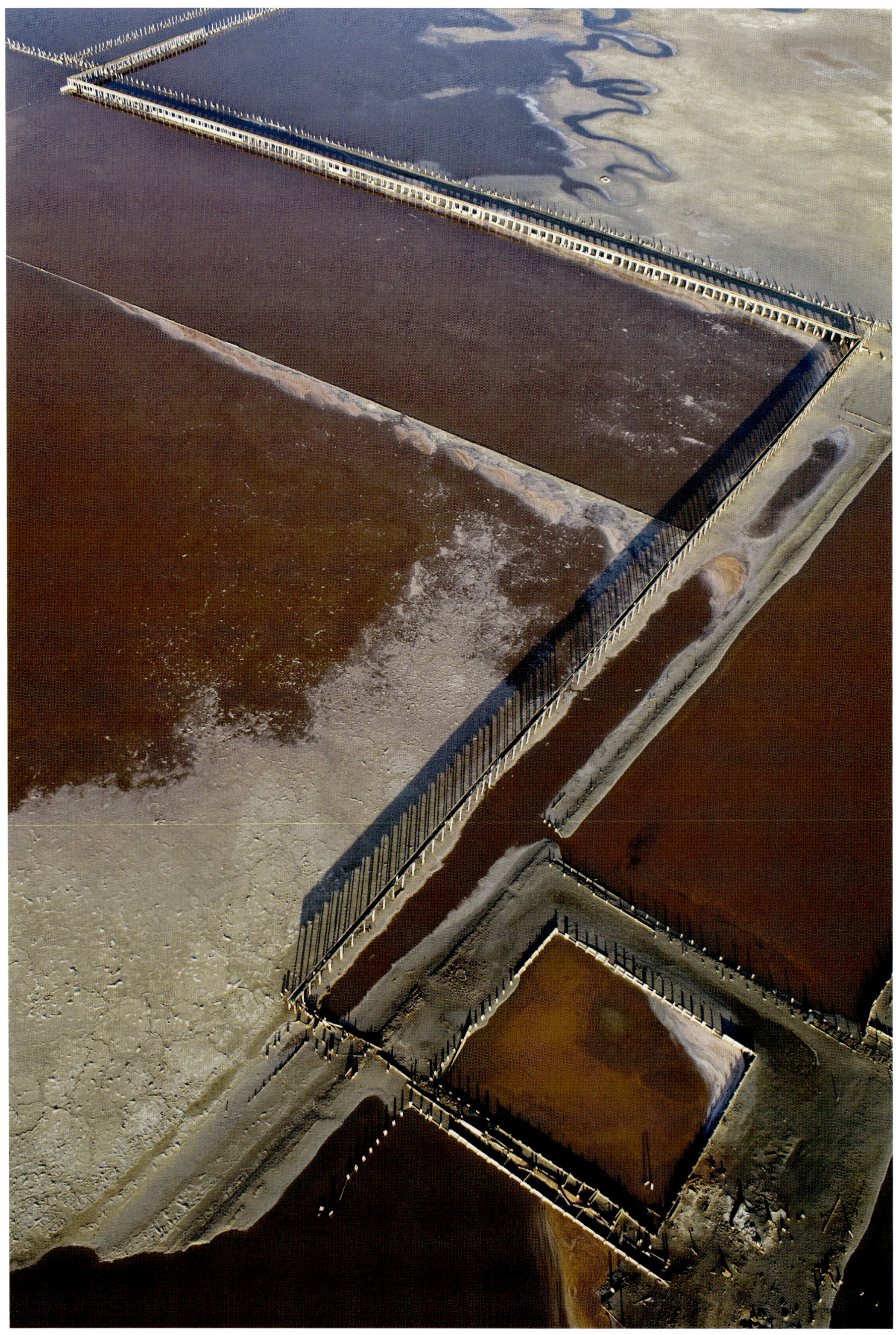

Plate 50. Artifacts near the old Whisby works (Salt Pond E13, November 2008)

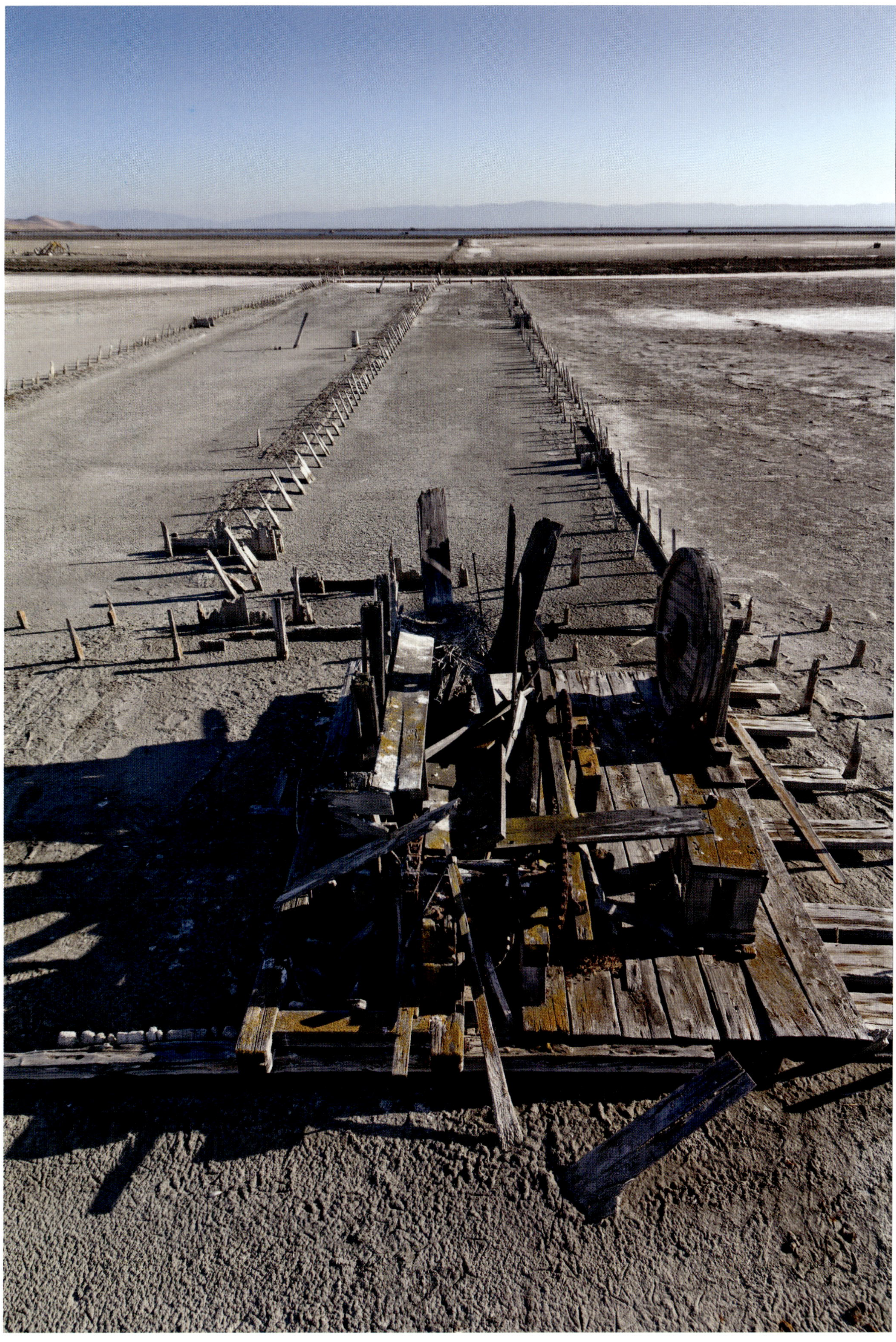

Plate 51. Ruins of early electric pump (Salt Pond E13, September 2010)

The salt business changed dramatically as the twentieth century arrived, and so began another transformation of the South Bay landscape. The business climate of the day favored consolidation; think of the board game Monopoly, which evolved from a game patented in 1904. Within a couple of decades, amid attempts to corner the salt market, corporate mergers rearranged over thirty salt companies into a half-dozen major players. At the same time, the amount of land in salt production increased substantially as corporate land speculators bundled property and shaped the policy of land reclamation districts. The larger companies invested corporate capital in new technologies to modernize their works, a strategy that leveraged economies of scale. With devices like the electric pump, siphons, and pipelines, salt-evaporation networks were recast as configurations of much larger ponds supplying far fewer processing plants. The new technologies allowed pond networks to span previous constraints in the landscape, such as larger marsh channels, and ultimately the bay itself (via an underwater brine pipeline). Levees became much more robust as construction shifted from hand labor to diesel-powered dredges. Mechanized harvesting arrived as purpose-built

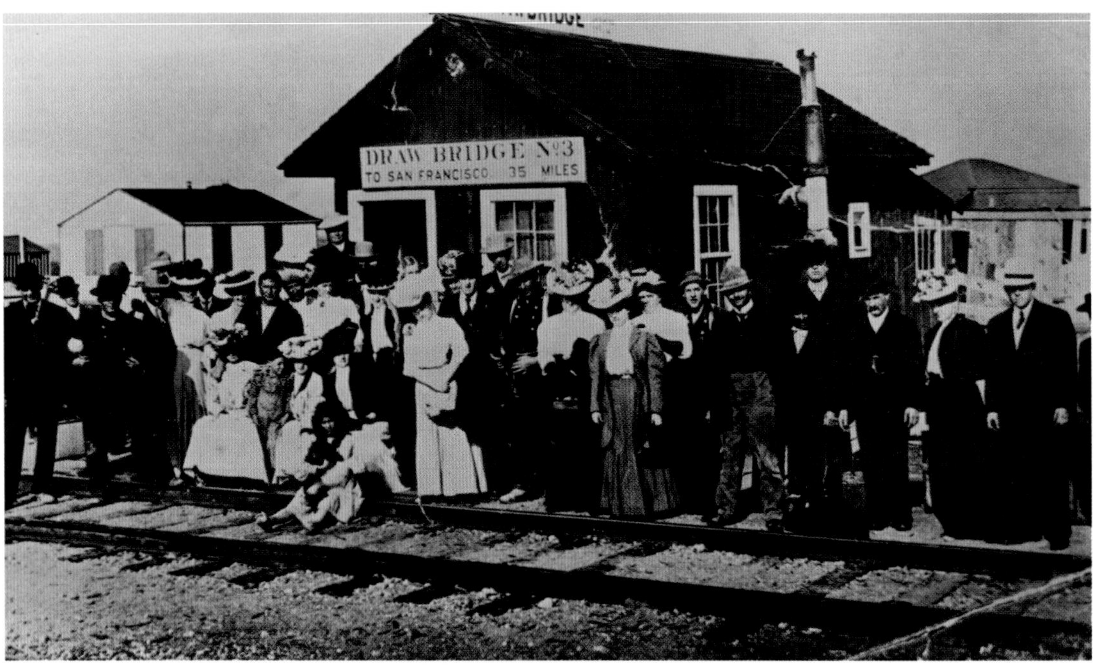

Waiting for the train at Drawbridge: the narrow-gauge South Pacific Coast Railroad provided the principal means of access to Drawbridge, serving the hamlet's residents and visiting duck hunters alike. Courtesy of Don Edwards San Francisco Bay National Wildlife Refuge

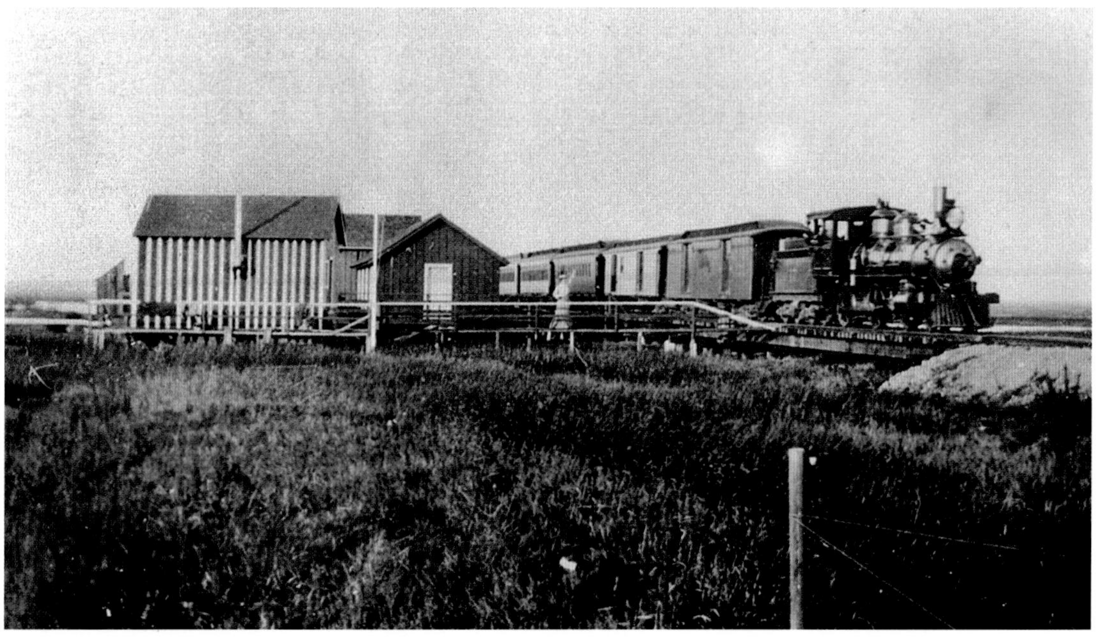

A South Pacific Coast Railroad narrow-gauge locomotive photographed at Drawbridge. Courtesy of Don Edwards San Francisco Bay National Wildlife Refuge

machines replaced men swinging picks, while miniature train systems retired wheelbarrows. And the salt pond landscape rearranged itself to accommodate this new order.

Efforts in the early twentieth century to find more profitable uses for the land, specifically conversion to dairy pasture, agriculture, and industrial sites, generally failed. This left salt making and the outdoor sports of hunting and fishing as the primary uses of the wetlands. As landholdings grew and ponds changed scale, the salt pond landscape extended into the South Bay near Alviso and along the western shore of the bay. This was an expansion that would continue into the 1950s. The ponds became so large that a single evaporator pond could submerge the footprint of one or more previous salt companies.

As in the game Monopoly, railroads assumed an important role. The first company to cross the bay was the South Pacific Coast Railroad, established in 1876. Within several years it had opened a narrow-gauge rail line stretching from Alameda to Santa Cruz. Their single track right-of-way crossed five miles of South Bay marsh in a straight line running from Newark to Alviso, a feat requiring two swing bridges, many trestles, and prodigious berming. The single track was widened to standard gauge after 1900 and is still in use today. Where saltworks had previously been tied to navigable marsh channels for shipping, the railroad created a new geography of transportation and access to a market of national scale. The boat landings began a steady decline.

The South Pacific Coast Railroad produced another interesting development in the South Bay. The swing bridges built across Warm Springs Slough (now Mud Creek) and Coyote Slough, roughly in the center of the marsh, required a tender to open the bridges for passing boats. George Mundershietz, the first bridge tender, found that life in the middle of the marsh offered agreeable access to prime hunting and fishing territory. His guests concurred and before long small shacks began to appear in the marsh along the railroad on Station Island, as it came to be known. Within a couple of decades the hamlet of Drawbridge was established. In 1906, Drawbridge had grown to about eighty cabins and two hotels in a settlement stretching along the mile or so of rail line on Station Island. While the south end was predominantly Catholic and the north end Protestant, the entire population seemed unified in a devotion to an outdoor life of hunting, fishing, boatbuilding, and the challenging enterprise of living on an isolated marsh. Drawbridge became a weekend destination for San Francisco hunters as well, earning a place as a regular stop on the railroad schedule.

The first bridge across the open waters of San Francisco Bay was built for the 1908 Dumbarton Cutoff Line. Established as a shortcut route for San Francisco–bound freight trains, this rail route paralleled the Hetch Hetchy pipeline out to Dumbarton Point. There the narrows of the bay have provided an attractive crossing point for many forms of infrastructure. The rail line required swing bridges for the crossing over the bay proper and over Newark Slough. These bridges are still in place although the rail line has been in disuse since 1982.

In photographing Dumbarton Point I have marveled at engineering that established an accurately pivoting rail bridge on foundations placed in a marsh slough. The area is also notable for the large expanse of original marsh surrounding the rail line and the neighboring Hetch Hetchy pipeline. While the Dumbarton Cutoff Line is quiet in the current day, the track running through Drawbridge bustles with traffic—Amtrak and the ACE line carrying passengers, as well as slow, rumbling freight trains. The trains roar through an abandoned Drawbridge, its cabins and outbuildings sinking into the marsh.

The mechanized transformation of the South Bay into a landscape of large salt ponds proceeded with vigor into the mid-twentieth century. By 1950, salt ponds had replaced most of the marsh that gave Drawbridge vitality. Untreated sewage and urban runoff tainted the wetlands that had escaped development. The South Bay became an inhospitable place. By many measures this could be considered a low point in the history of the landscape.

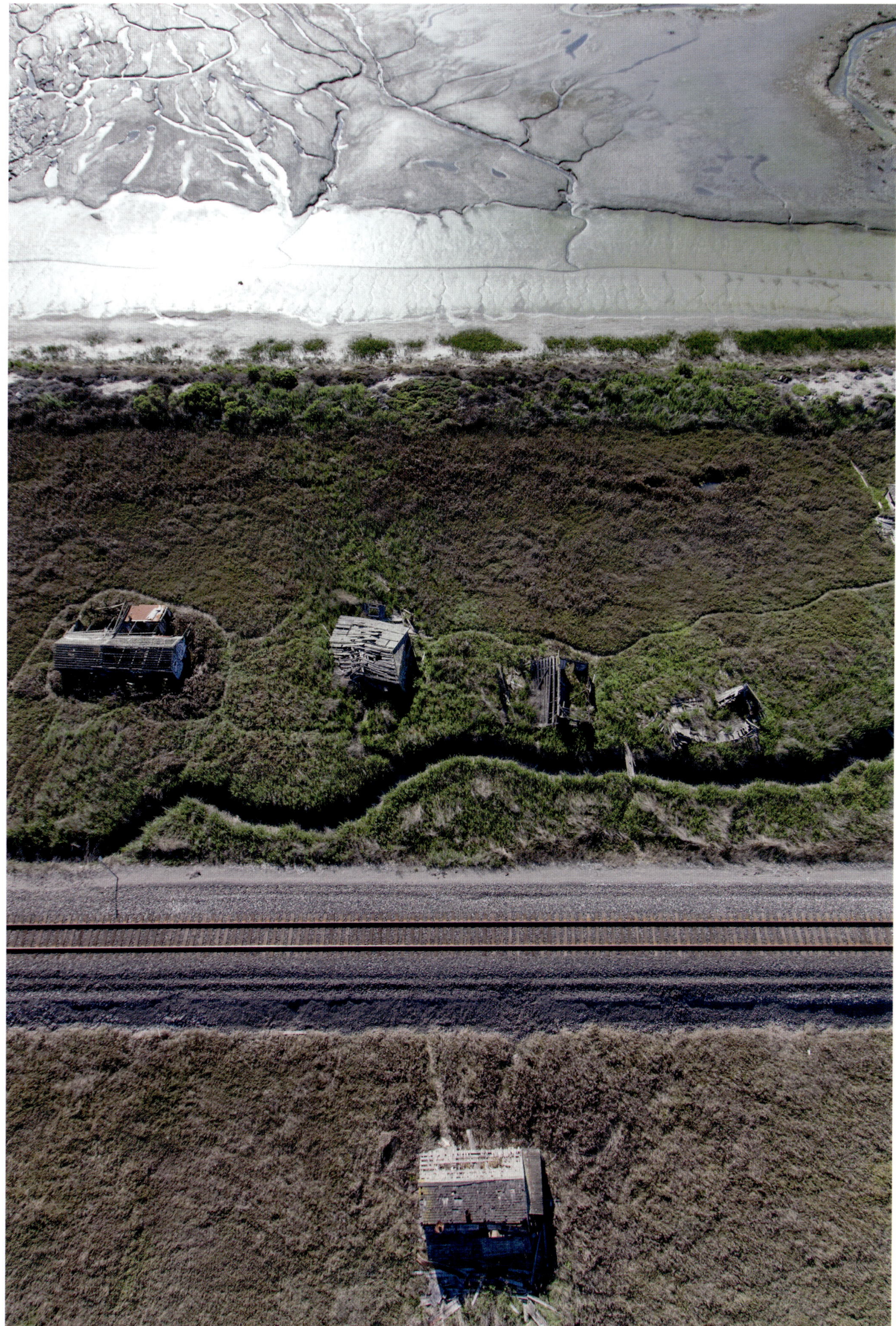

Plate 52. Railroad grade and Salt Pond A21 (Drawbridge, April 2012)

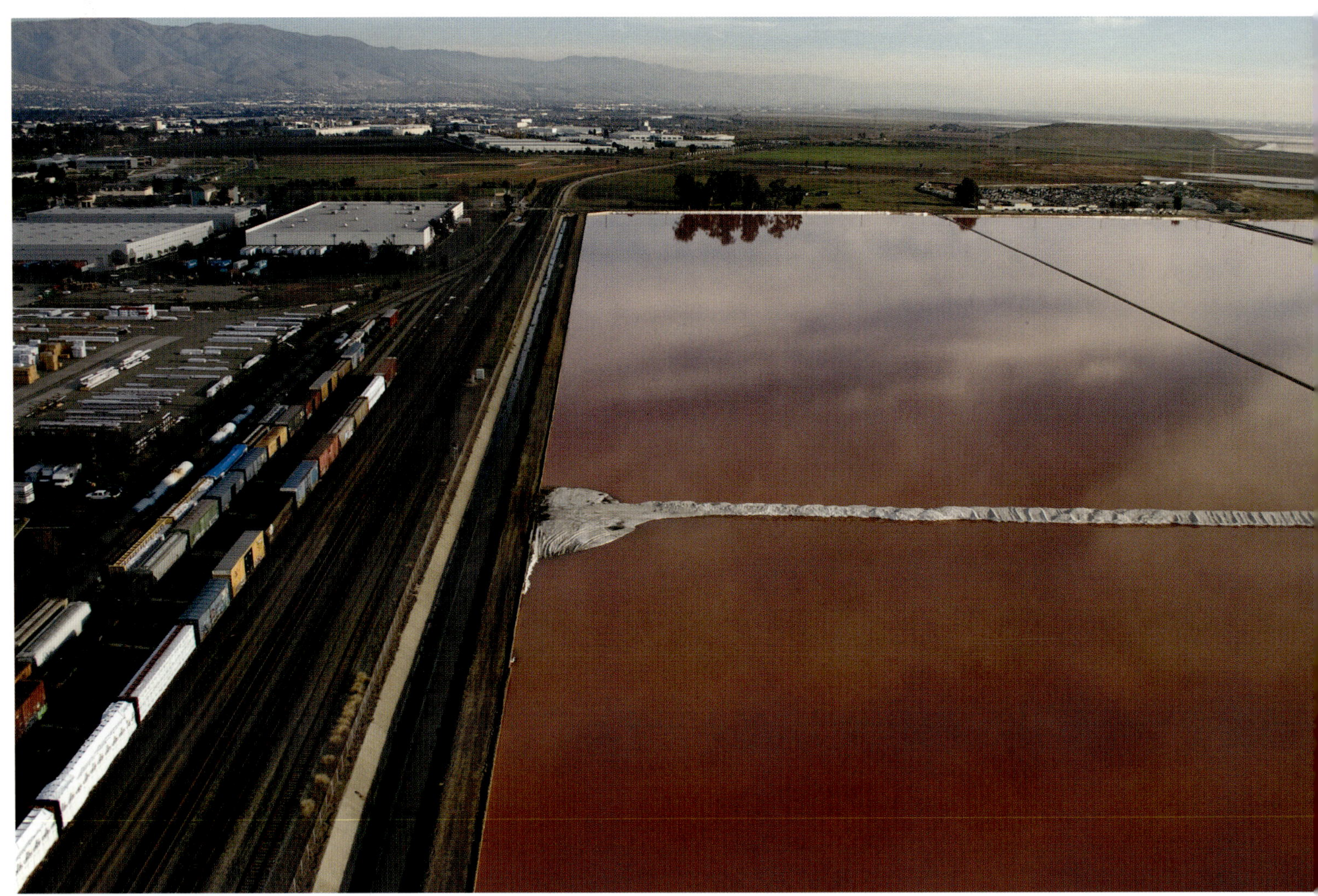

Plate 53. Rail corridor (Newark Crystallizer Beds, January 2008)

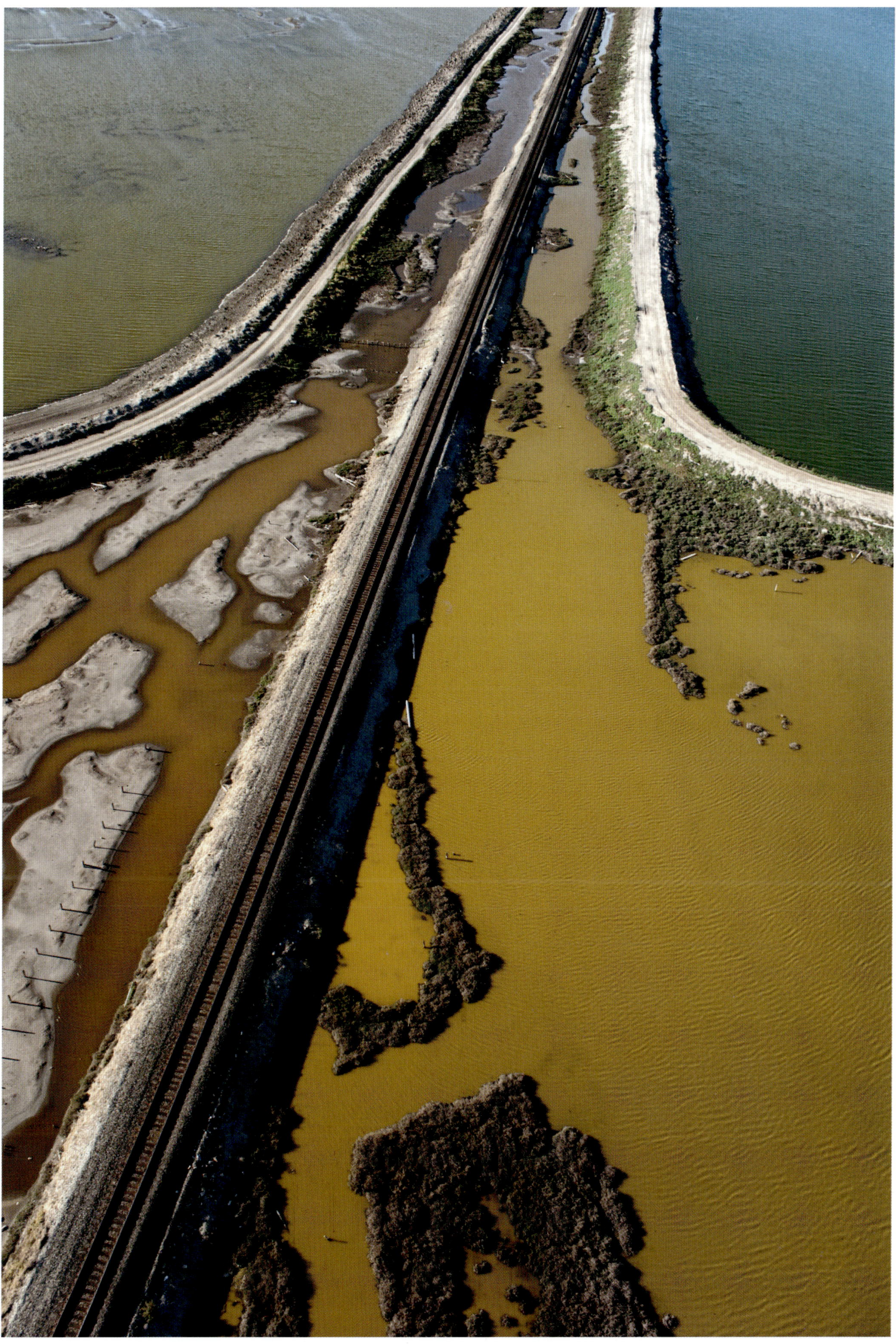

Plate 54. Railroad grade flanked by ponds (Union Pacific Railroad corridor, April 2009)

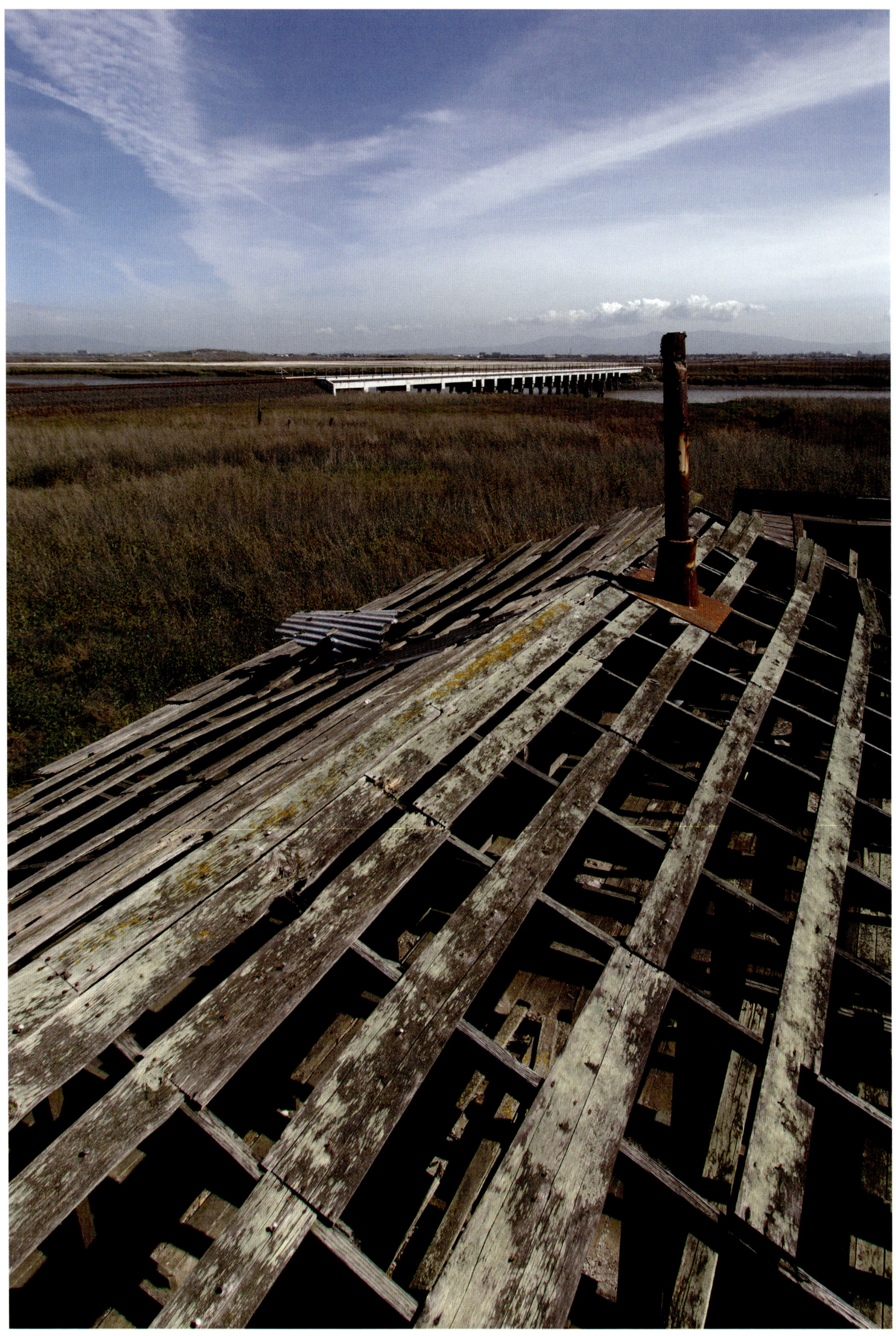

Plate 55. Shack roof (Drawbridge, April 2012)

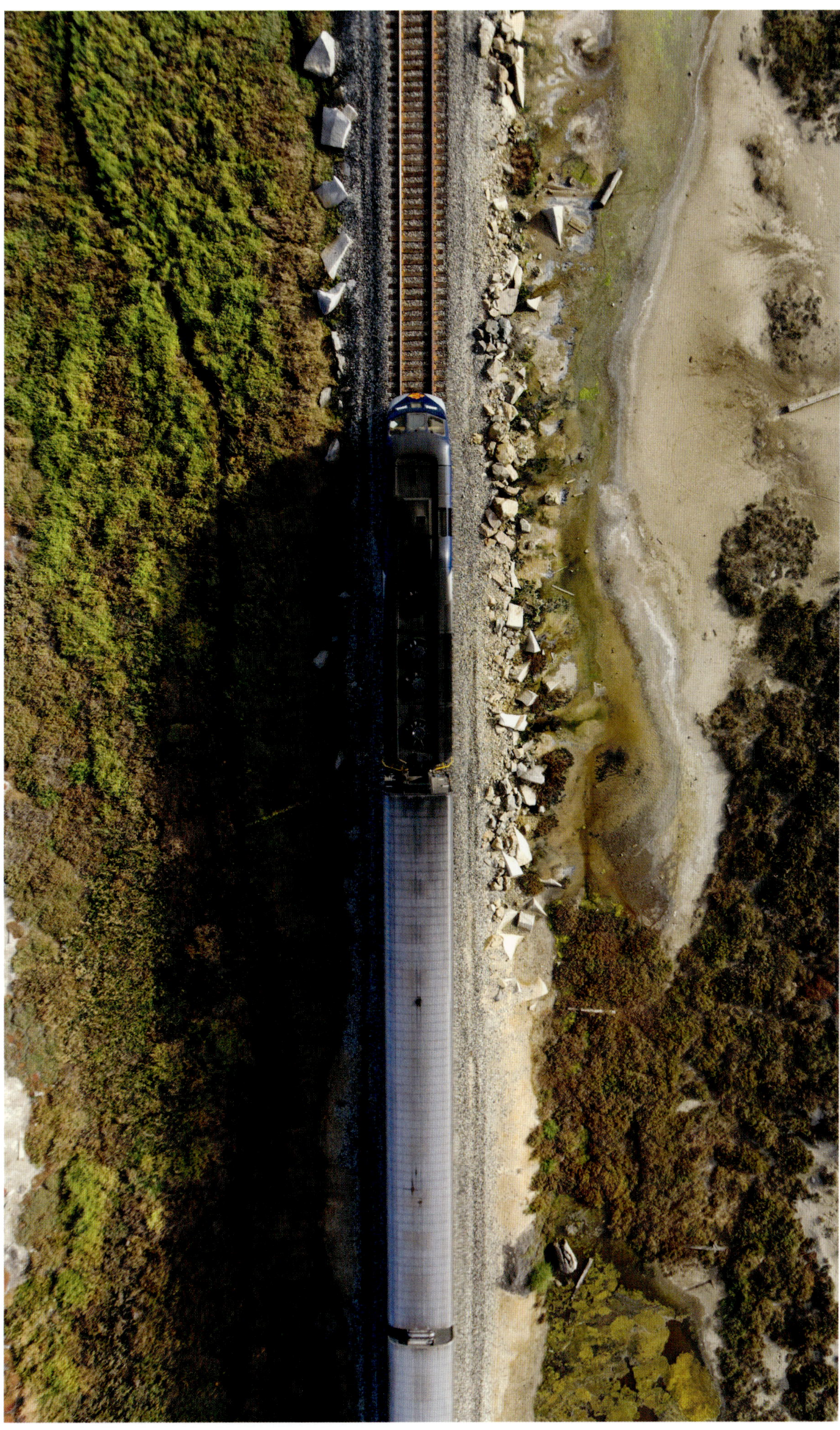

Plate 56. Amtrak at high velocity (Salt Pond A12, September 2011)

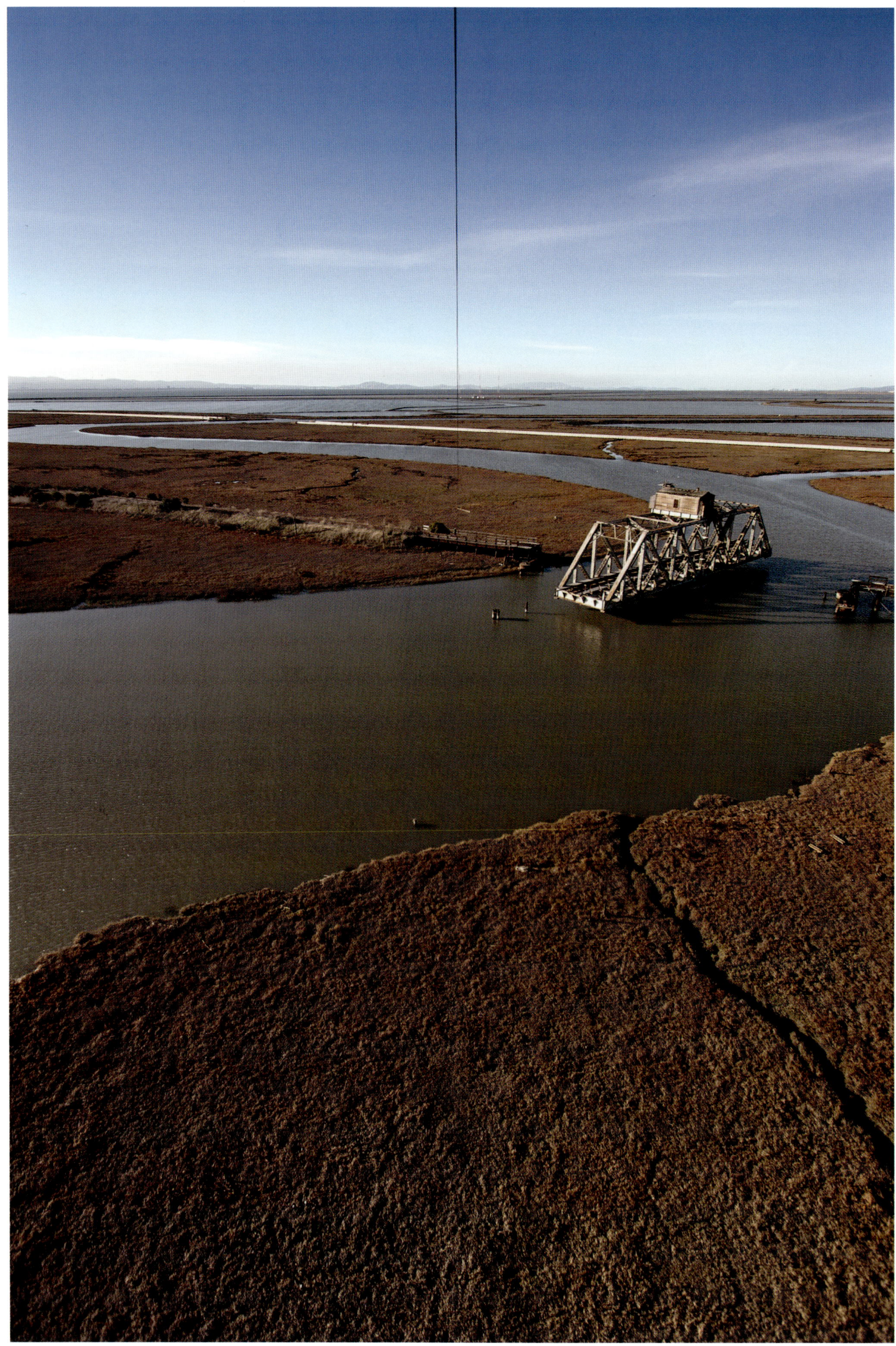

Plate 57. Bridge and surrounding marsh (Newark Slough, November 2009)

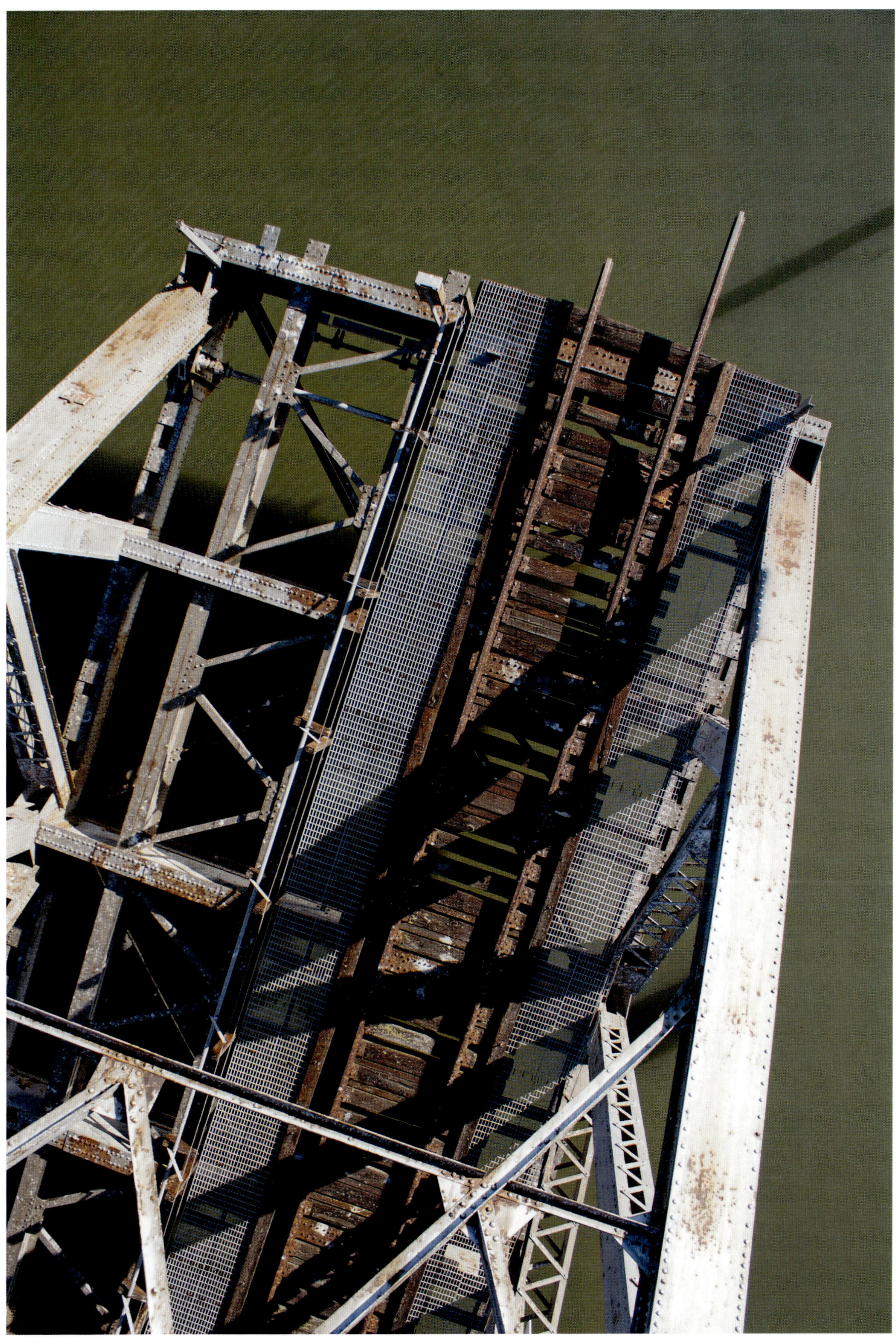

Plate 58. Detail of swing bridge structure and tracks (Newark Slough, November 2009)

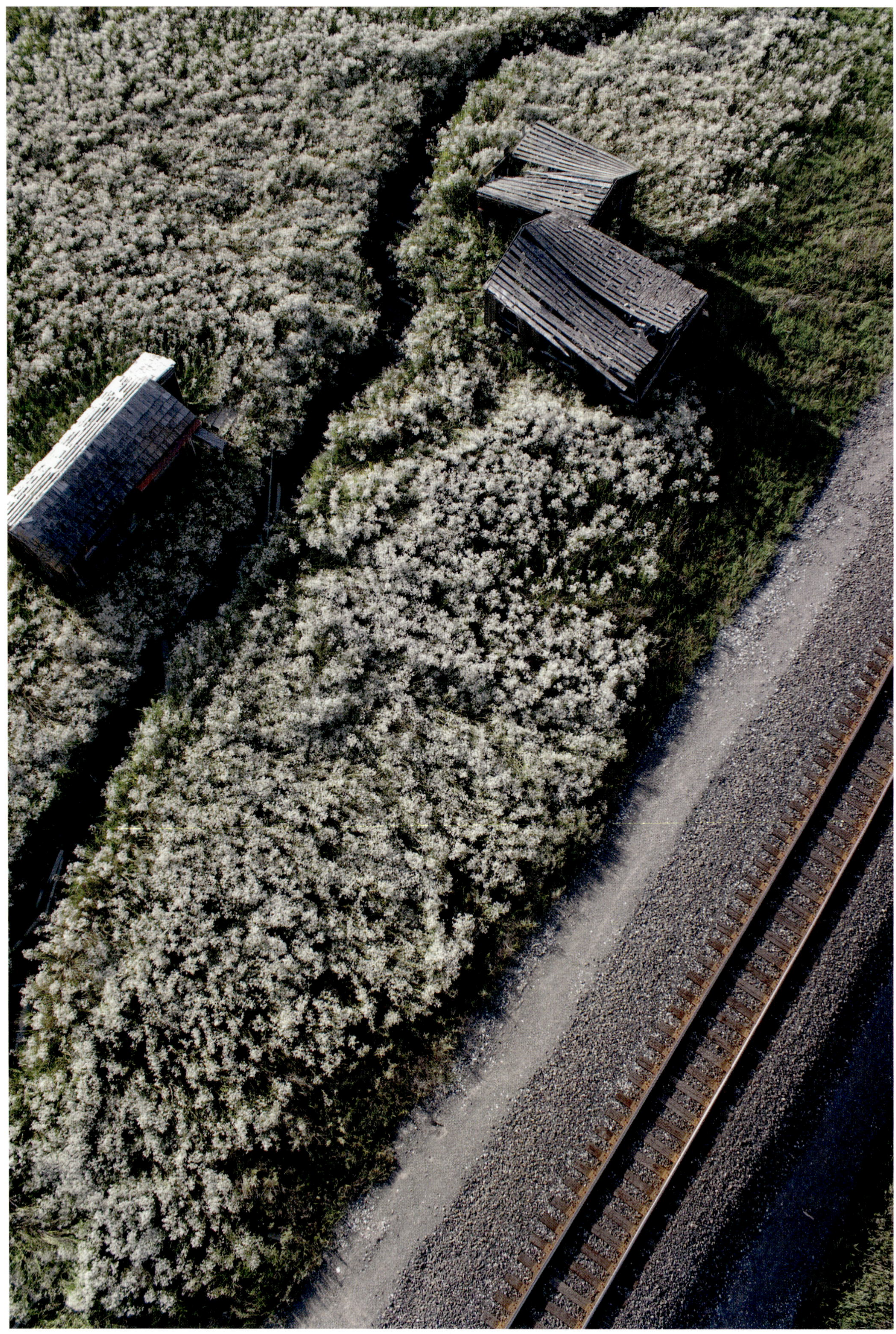

Plate 59. Returning to the earth (Drawbridge, May 2010)

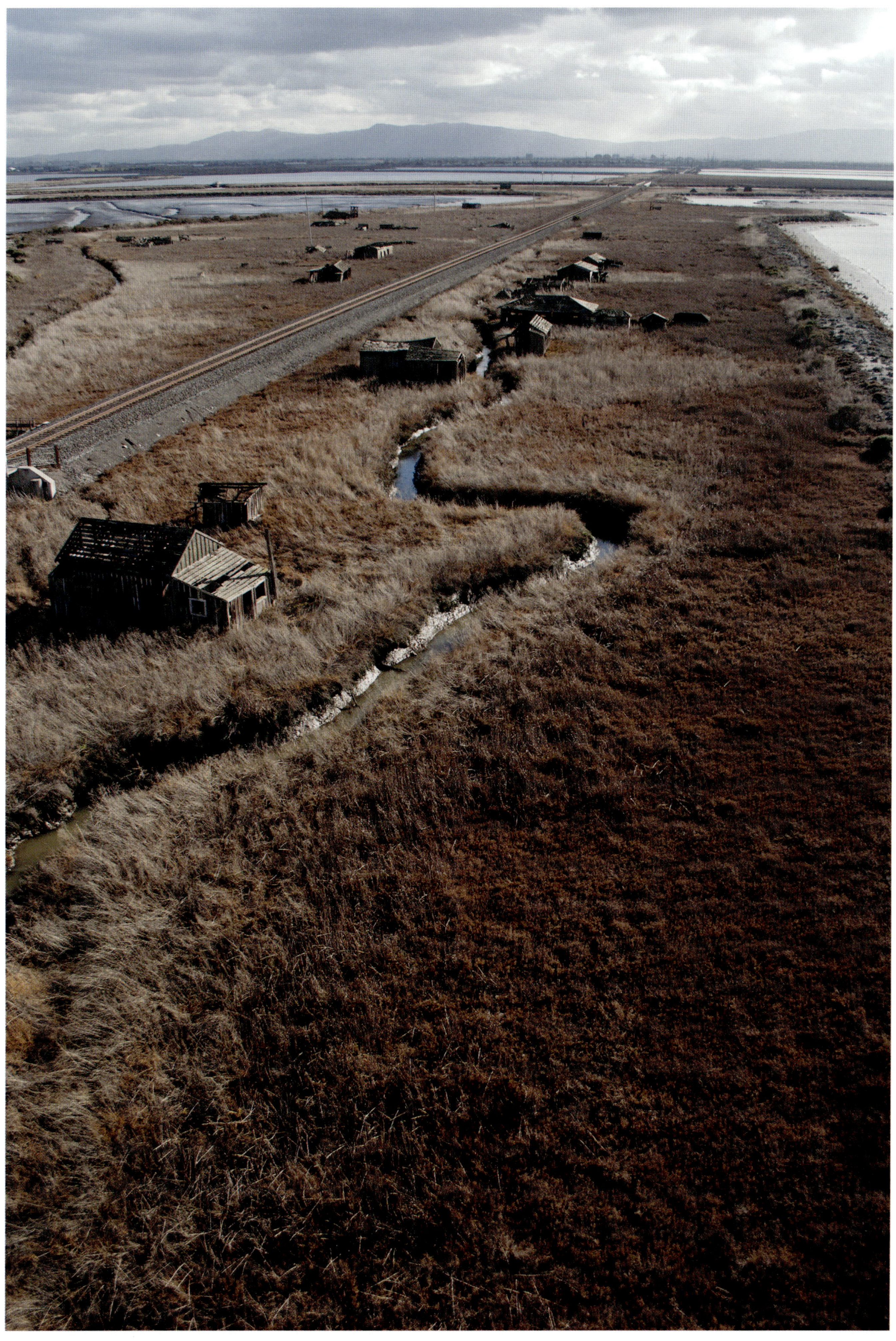

Plate 60. Abandoned houses (Drawbridge, January 2008)

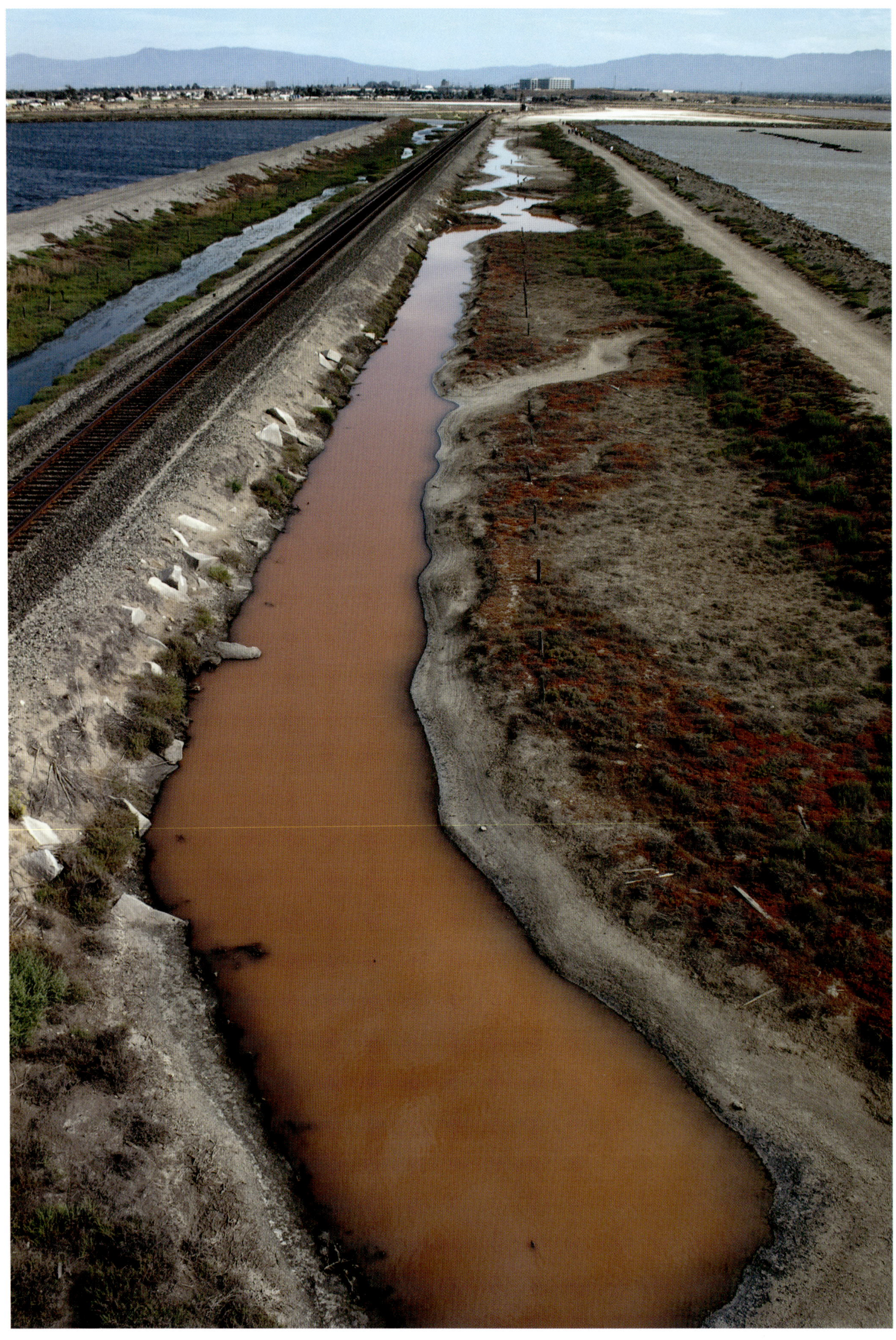

Plate 61. Trackside ditch (Union Pacific Railroad corridor, July 2009)

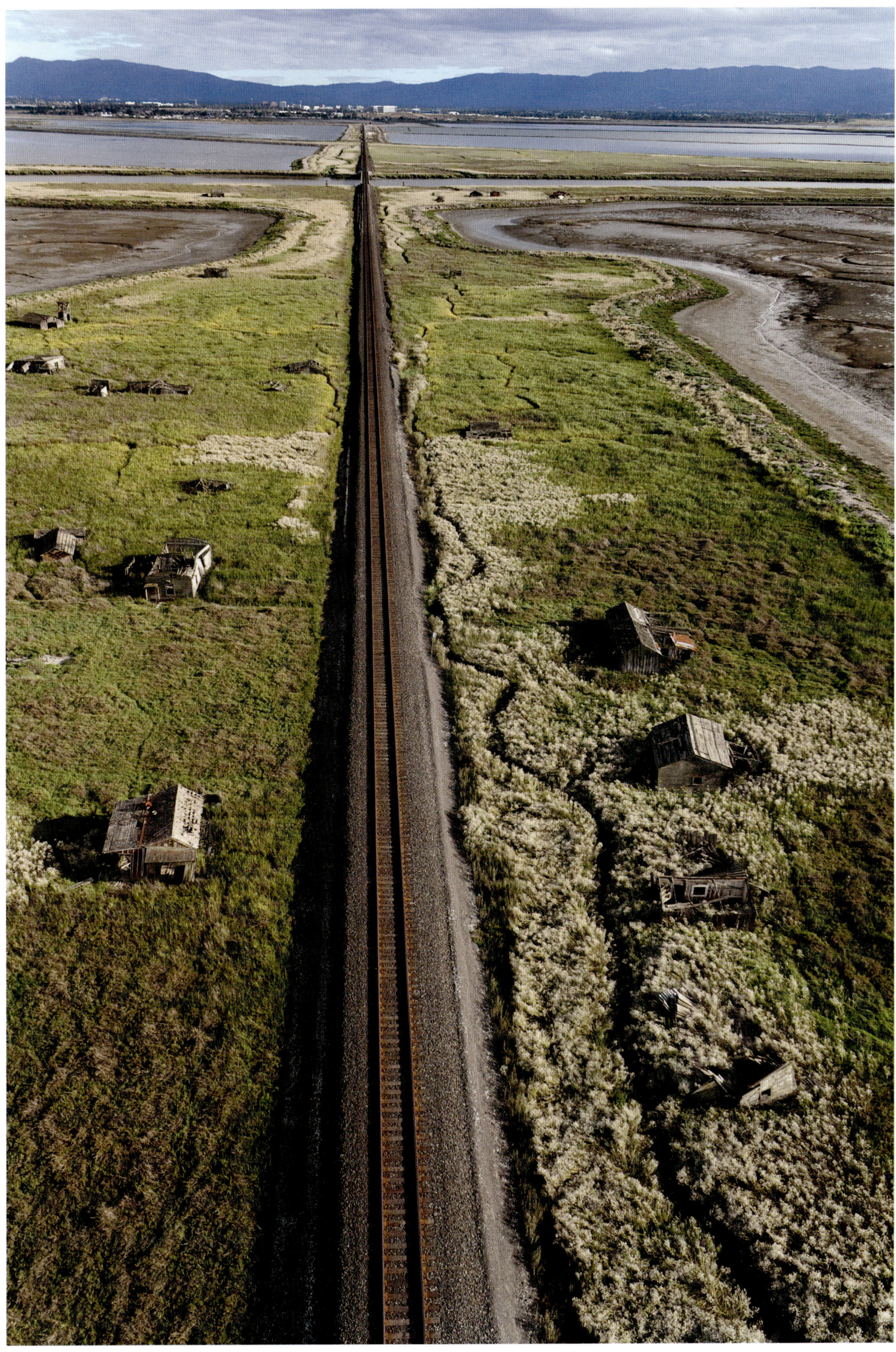

Plate 62. Down the tracks toward Alviso (Drawbridge, May 2010)

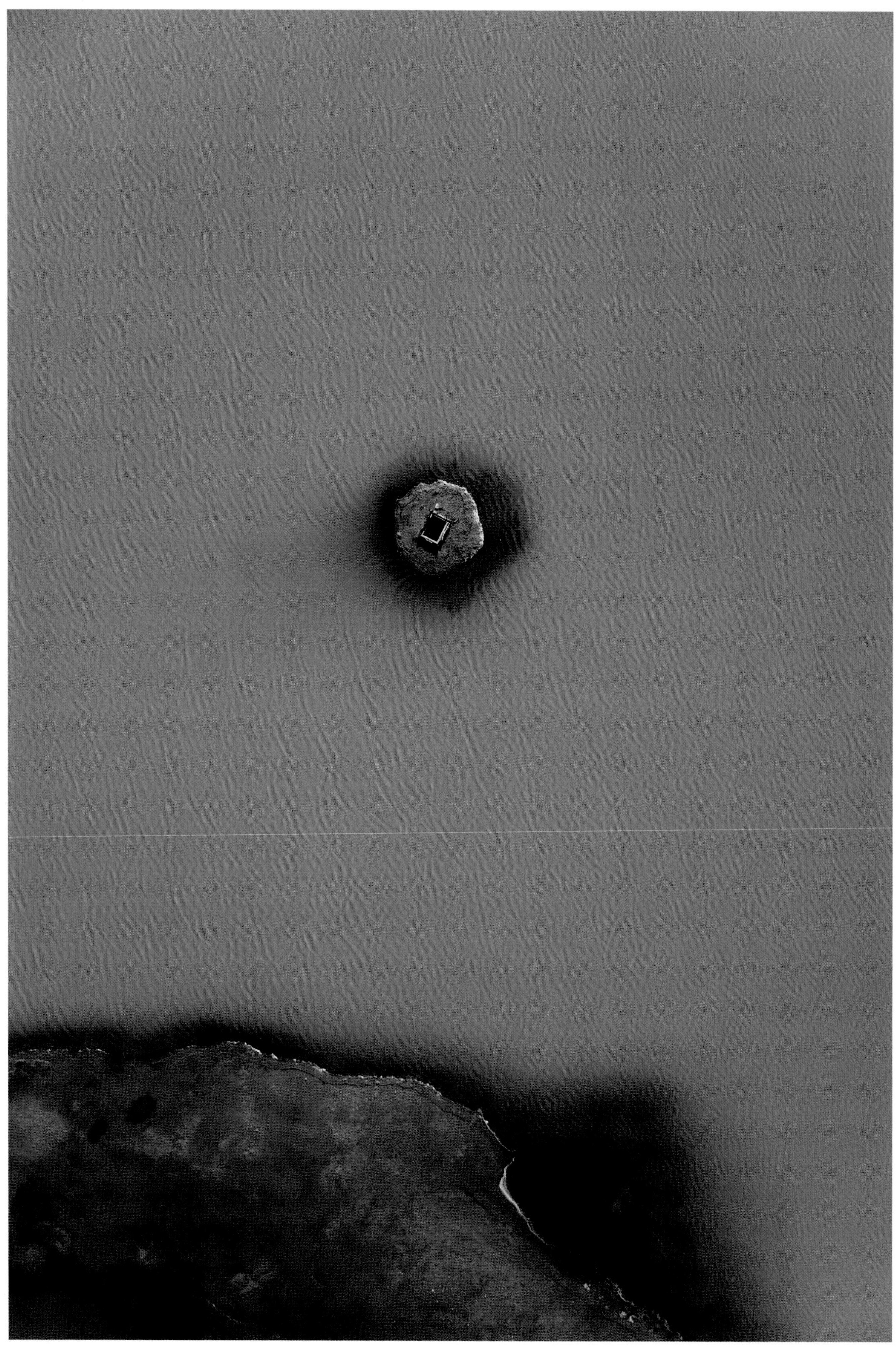

Plate 63. Dugout blind in small mound (Salt Pond E6B, November 2009)

5: CURIOUS MEASURES

 offer features to guide our understanding of the view below. Even in abstract images, familiar items can convey, in their size and arrangement, clues that help us interpret the lay of the land in question. For instance, regularly spaced crop rows establish a sense of scale in images of farmland, and the alignment of cultivated rows with topographic contours further sets the scene. The particular geometry of creosote plants in the desert presents a pattern strongly wed to that place, one established by the basic mechanisms of life in difficult circumstances. Compositionally, it is often fun to avoid overt clues in aerial photographs, to seek images of seemingly indeterminate scale or dominated by visual patterns that beg a second look. The South Bay offers many opportunities for such abstract play. Yet even the most abstract of my images contain measures of the landscape—clues regarding scale and circumstance.

The dendritic patterns of dormant marsh channels became a familiar measure in my early South Bay photographs. As I gained experience in the landscape other measures emerged; curious features that, once understood, supported a reading of the place. Some of these measures were evident in plain sight as I walked the South Bay levees, while others lurked below the surface of shallow brine. Most easily seen were the levees proper—straightforward, dominant elements in this landscape. Contemporary South Bay salt ponds share a basic schema in which a large shallow plateau, the former marsh bottom, is covered by one to three feet of brine. A perimeter ditch six to ten feet deep surrounds the plateau. Mud was dredged from this "borrow ditch" to create and maintain the enclosing levee.

Seen from a distance, variations in levee layout and shape offer clues about former features in the landscape and the history of local interventions. Since 1936 a single dredge, the *Mallard II*, has had principal responsibility for levee construction and maintenance in the South Bay. A closer view of South Bay levees finds marks made by the dredge in the process of their construction, ubiquitous artifacts that provide a spatial and temporal datum. Similar in scale but quite different in origin are duck blinds, a feature dotting the landscape in patterns tuned to the carrying power of a shotgun. The South Bay has a storied history of duck hunting, a collateral use of the wetlands that continues today. All told I have become quite fond of finding these measures of the landscape, elements that belong to their place and its history while simultaneously lending scale and meaning to my images.

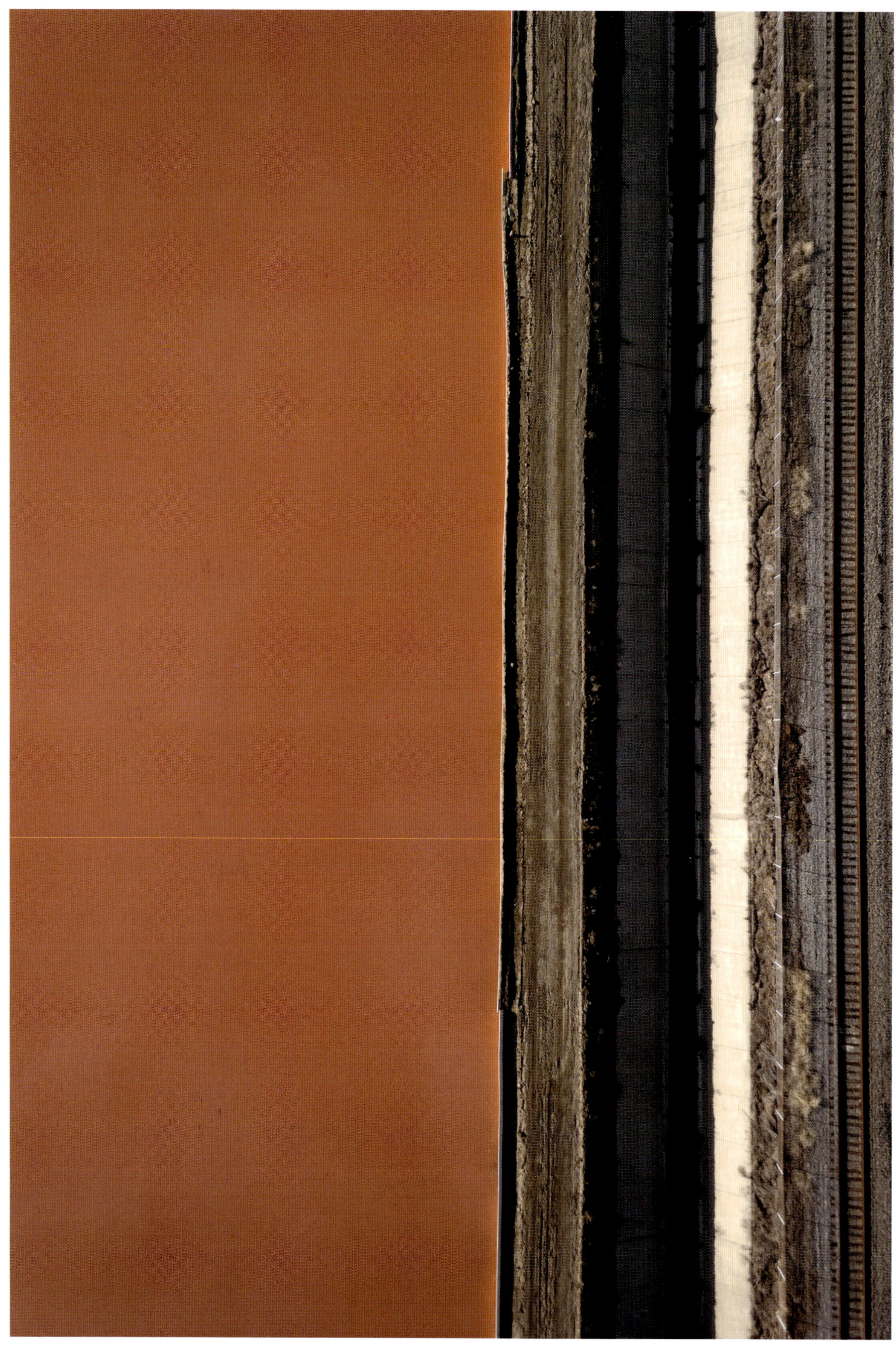

Plate 64. Compressed infrastructure at Newark edge (Cargill Crystallizer Beds, January 2008)

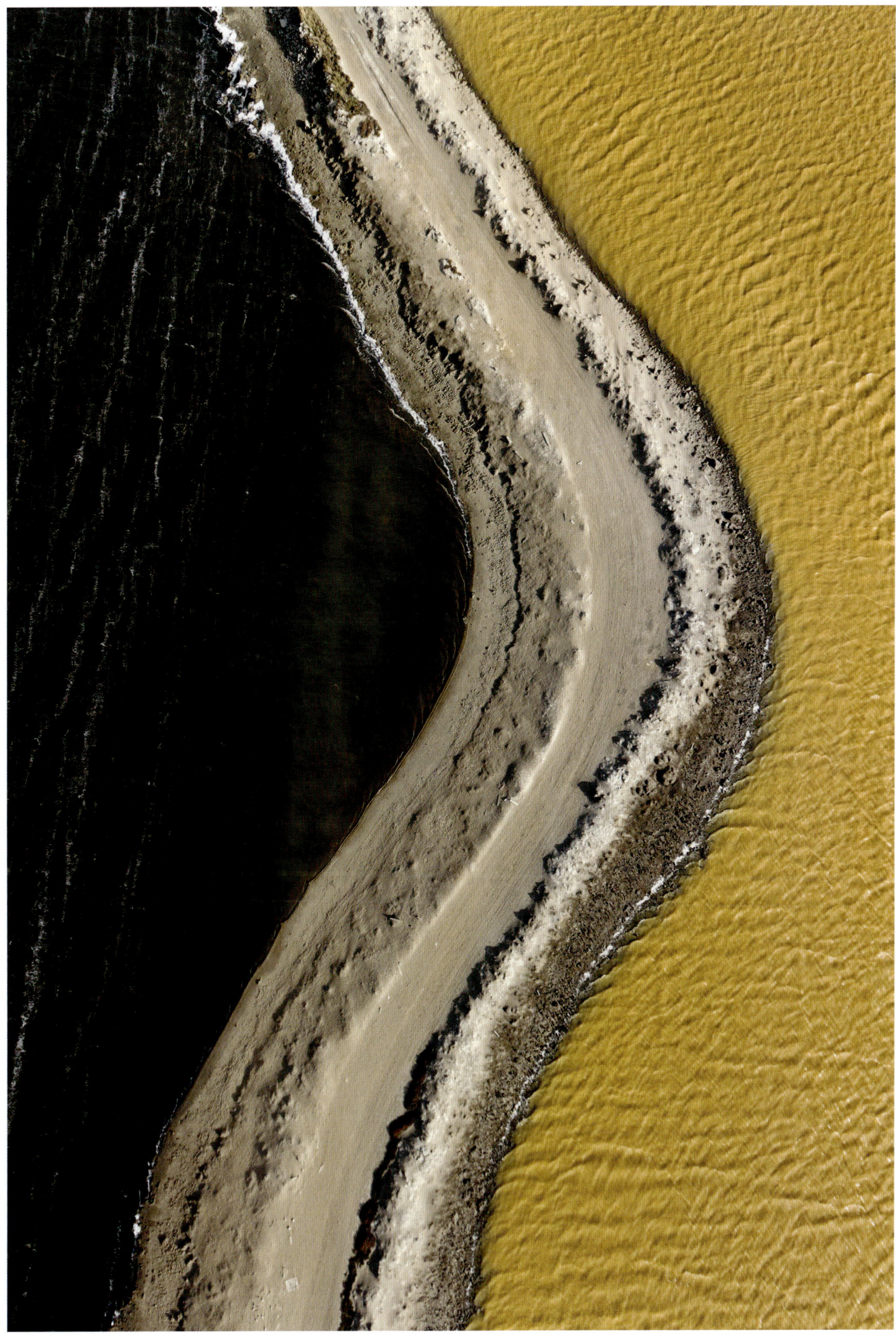

Plate 65. Windblown erosion (Salt Ponds A13 and A15, September 2011)

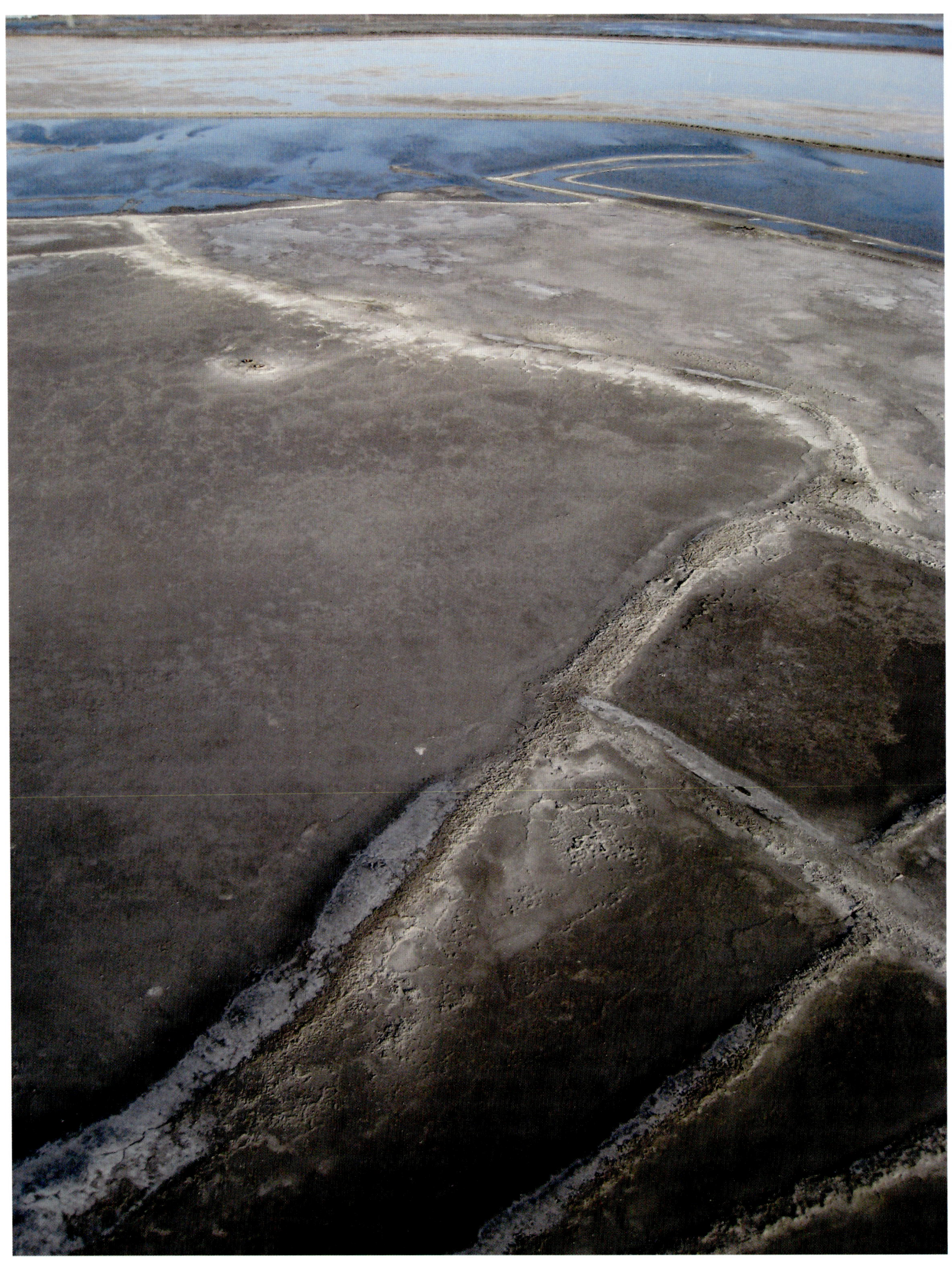

Plate 66. Old, irregular levees (Salt Pond A22, April 2007)

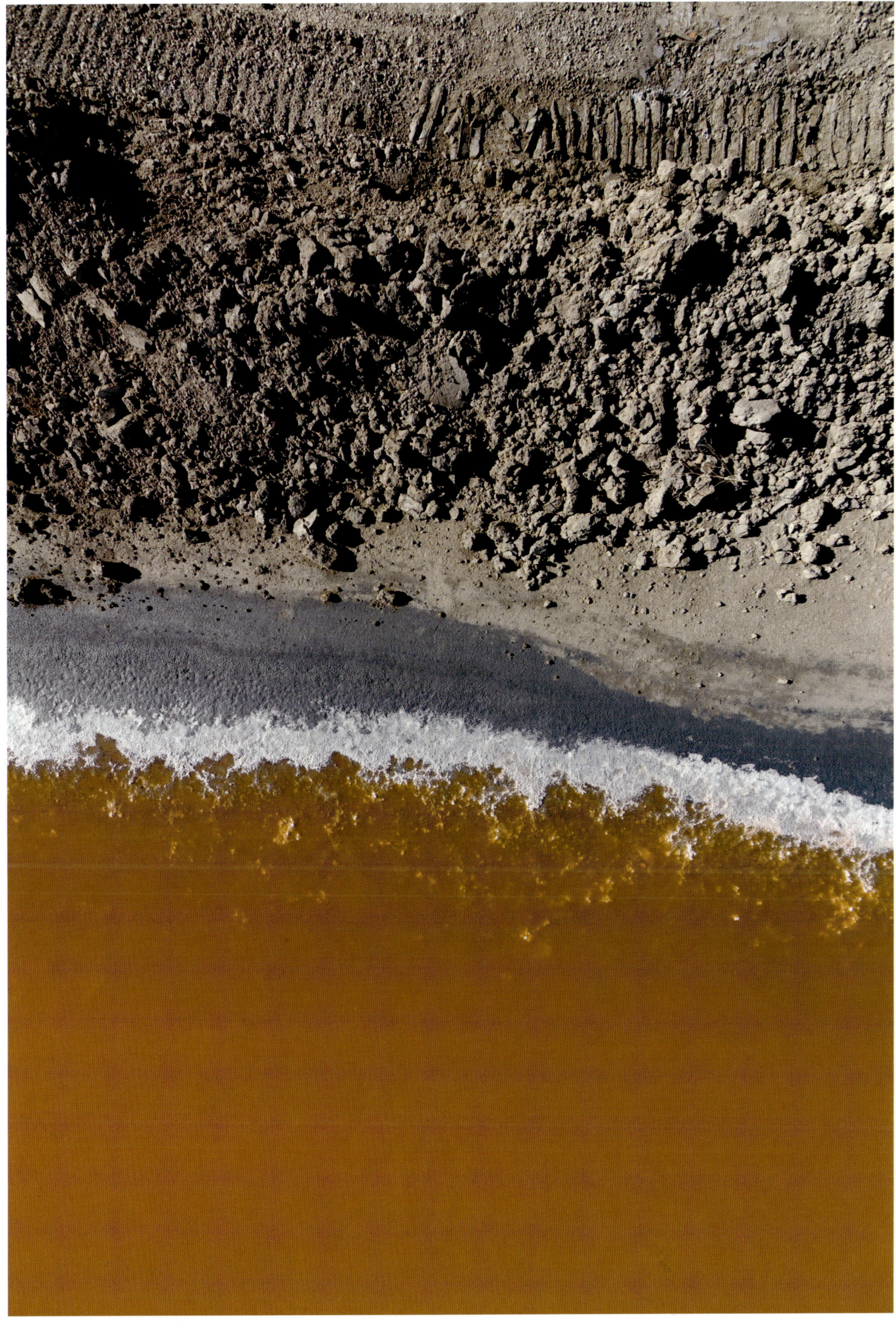

Plate 67. New levee from imported fill (Salt Pond E13, August 2011)

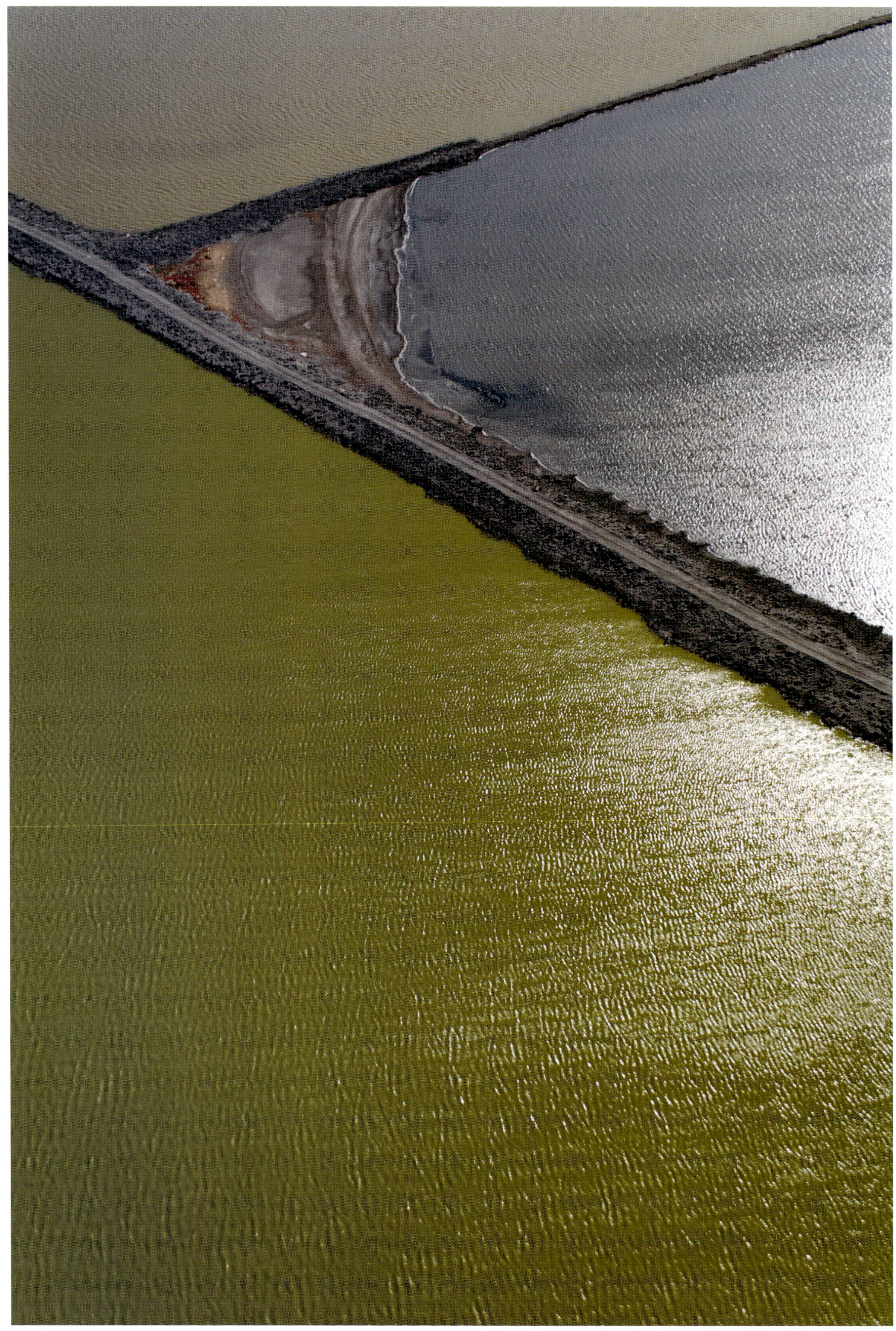

Plate 68. Surveyor's levees (Salt Ponds M1, M2, and M3, July 2008)

By 1936, the Leslie Salt Company had acquired all other salt operations in the Bay Area except the Oliver Brothers Salt Company and was the dominant, and eventually sole, Bay Area producer of salt. Leslie then set about revising the South Bay landscape to become a unified and expanded network of much larger evaporator ponds. Gone were the days of modest levees built by hand or with animal power. The large ponds required more substantial levees and toward that end Leslie commissioned the *Mallard II*, a long-boom, gravity-swing, grab dredge purpose-built for levee construction. Despite its twentieth-century credentials, the wooden-hulled *Mallard II* conveys a sense of Victorian engineering. While huge, it is an intriguingly simple machine operated by mechanical winches, which are controlled by levers to repeat the same dance pattern over and over. The dredge bucket, hanging from the end of a 103-foot wood boom, is dropped with jaws wide open into borrow-ditch mud. The closing wire pulls to close the bucket and grab a load of mud, then the hoisting wire lifts the bucket from the ditch. Once aloft the bucket swings over the levee, where its mud cargo is released in just the right place, and the boom returns to the borrow ditch. Repeat ad infinitum.

I get genuinely excited when I come across the *Mallard II* on a hike. Now seventy-five years old and operated by Cargill Salt, the modernized dredge is still hard at work maintaining South Bay levees in a steady struggle against the effects of erosion, subsidence, and soil compaction. The *Mallard II* meets the gaze as an ungainly leviathan, yet it roams widely in the South Bay. I have found it working on levees from Alviso to Eden Landing. The dredge moves ever so slowly by grabbing the pond bottom and pulling itself along. It drops spuds, something like temporary pilings, to anchor itself for work and creates little circles of levee, dredge locks, as points of entry or egress when moving between ponds. On a good day the *Mallard II* renews, with a precision born of endless practice, about one hundred lineal feet of levee. I marvel that

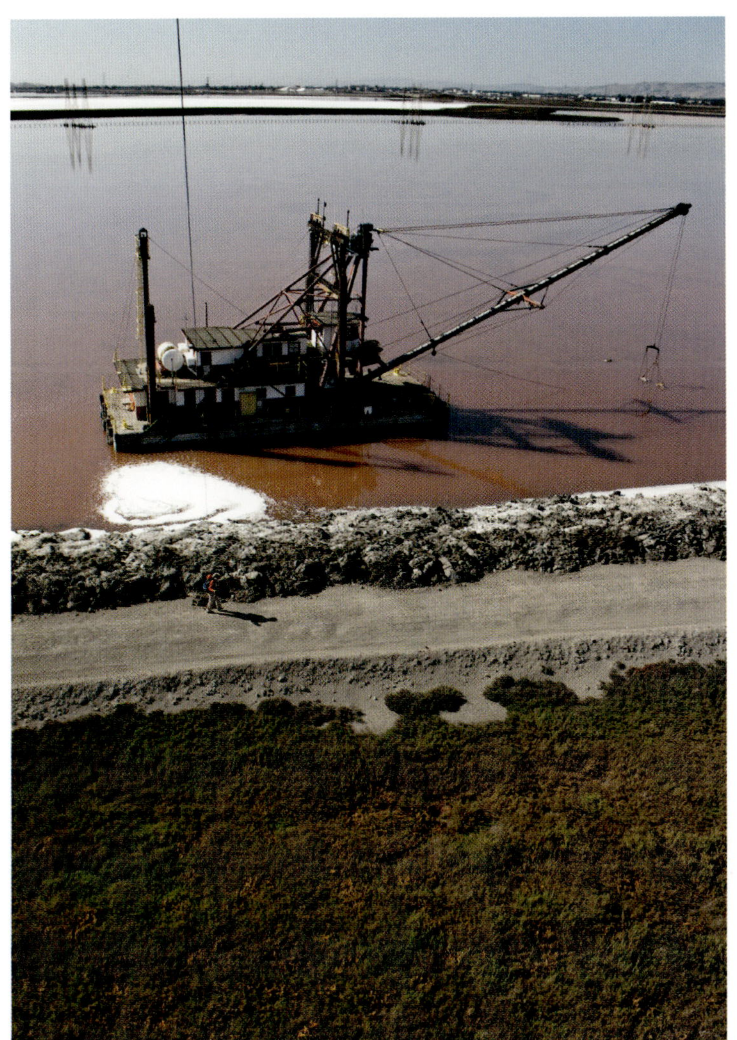

The *Mallard II* working in the M-Ponds north of Drawbridge. Freshly deposited mud is evident on the pond side of the levee coming up to about my shoulder height (I am visible in the left center of the image).

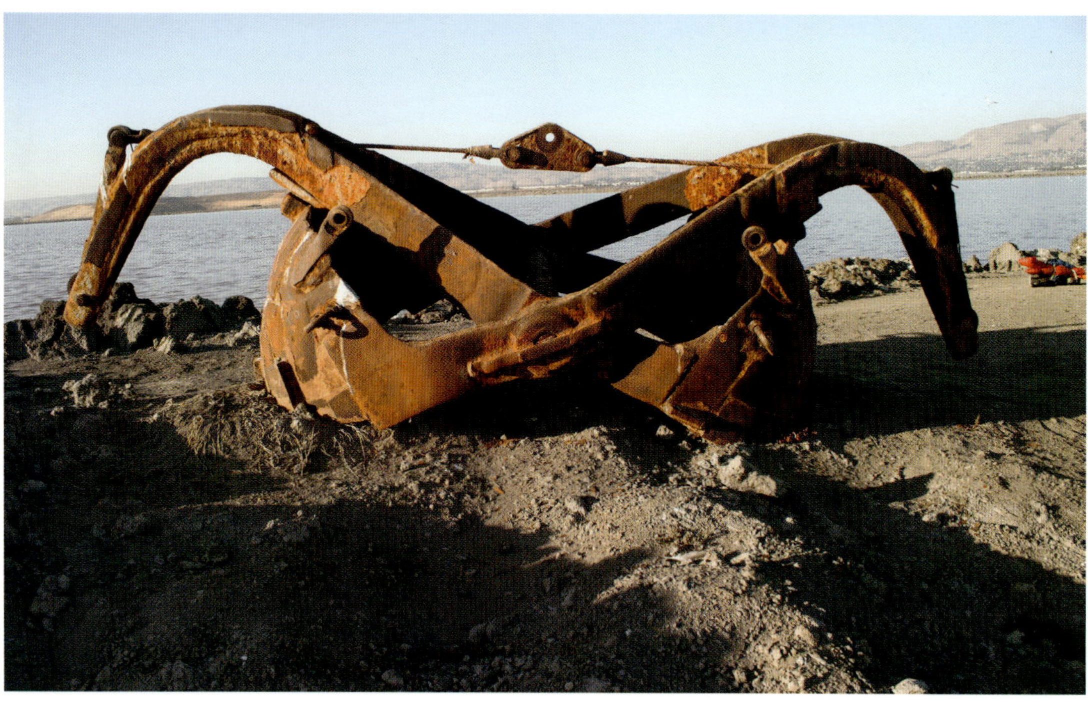

A dredge bucket from the *Mallard II*, encountered on the levee of Salt Pond M4 one fine Sunday—quite exciting.

the skilled crews in the lever house of this one machine, primitive by today's standard, created virtually all of today's South Bay salt-pond landscape. The *Mallard II*'s role has an even larger reach since the *Mallard II*'s levees, created for the sole purpose of salt making, have become an ad hoc means of keeping bay waters out of our subsided Silicon Valley and, through the graces of Cargill Salt, the dredge serves as an instrument for wetland restoration.

While taking photographs in the South Bay I generally walk the levees. There to the side of each levee I can see the familiar borrow ditch, the levee's inverted doppelgänger. This borrow ditch is the source of mud mechanically excavated to make, and subsequently maintain, the adjacent levee. Much less evident from the ground, since they are often underwater, are marks made by the *Mallard II*'s dredge bucket while constructing the levee. The dredging process leaves very particular divots along the edge of a borrow ditch, created when the clamshell bucket grabs its load. The marks have fresh, straight edges when new, then soften and fade over the years due to erosion and sediment. From the air they leave an interesting record of levee construction. In my photographs they provide a measure of scale, since the mark left time and time again always has the nominal four-foot width of the dredge bucket. In high photographs the dredge marks are tiny and in low images they loom large. The softness of the marks, and some are barely discernable, provides a measure of time elapsed since the levee was last maintained. My photographs containing dredge marks can be pleasingly abstract while offering clues about scale and circumstance—most satisfying.

As I accumulated photographs of levees I started to ponder why some were straight while others were curved. I slowly realized, and why "slowly" I do not know, that the curved levees delineated, like lines on a map, locations once occupied by large marsh channels. Gazing across the salt ponds placed cheek to jowl in the current-day landscape, it stretches the imagination to think of this place as once organized around a series of now vanished channels. But here the memory of these channels remains in curved sweeps of levee. And if the curved levees represent former natural features, straight levees are largely the mark of surveyors, lines drawn in service of ownership transactions or the flow of brine for salt making.

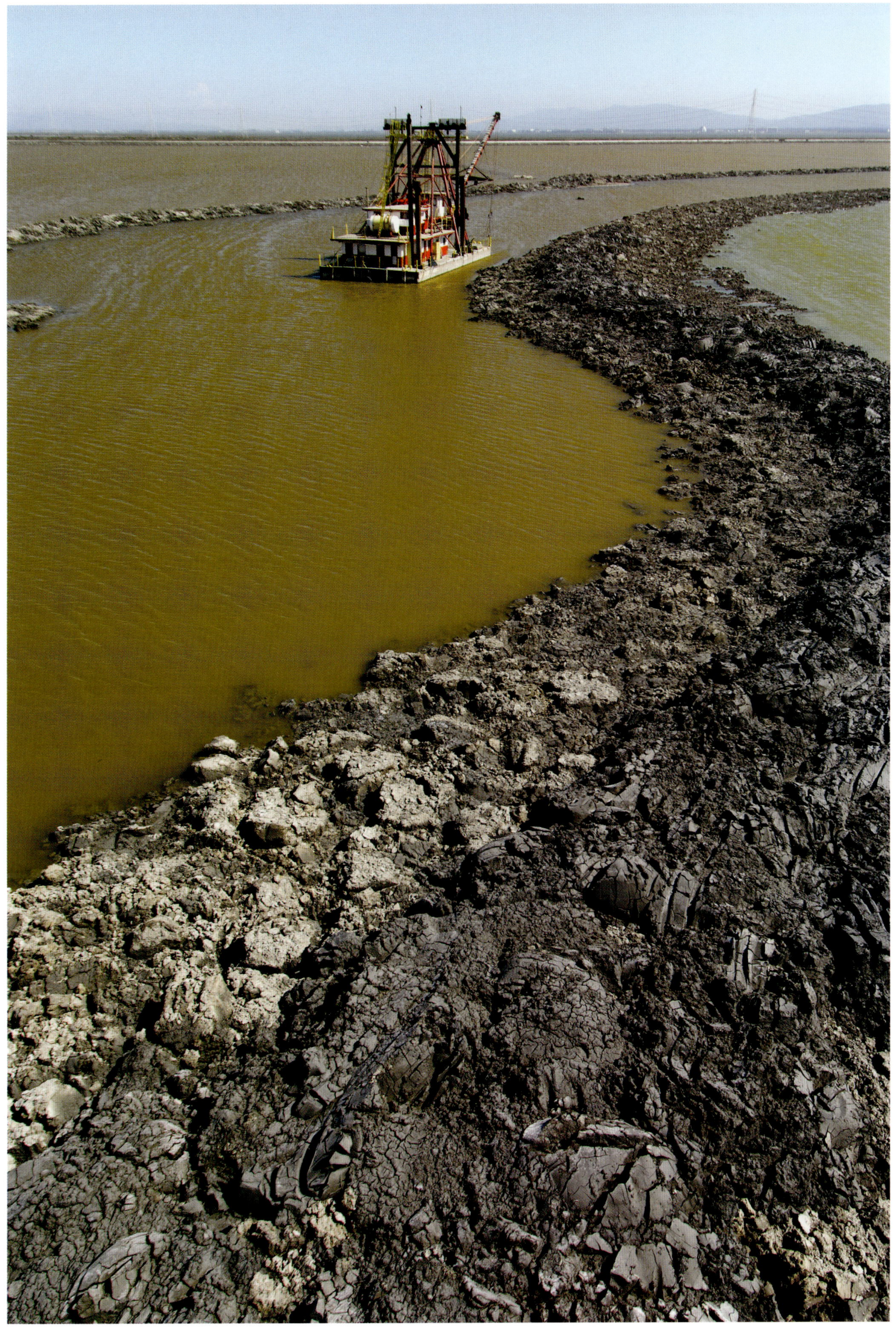

Plate 69. *Mallard II* renewing levee (Salt Pond N2, April 2012)

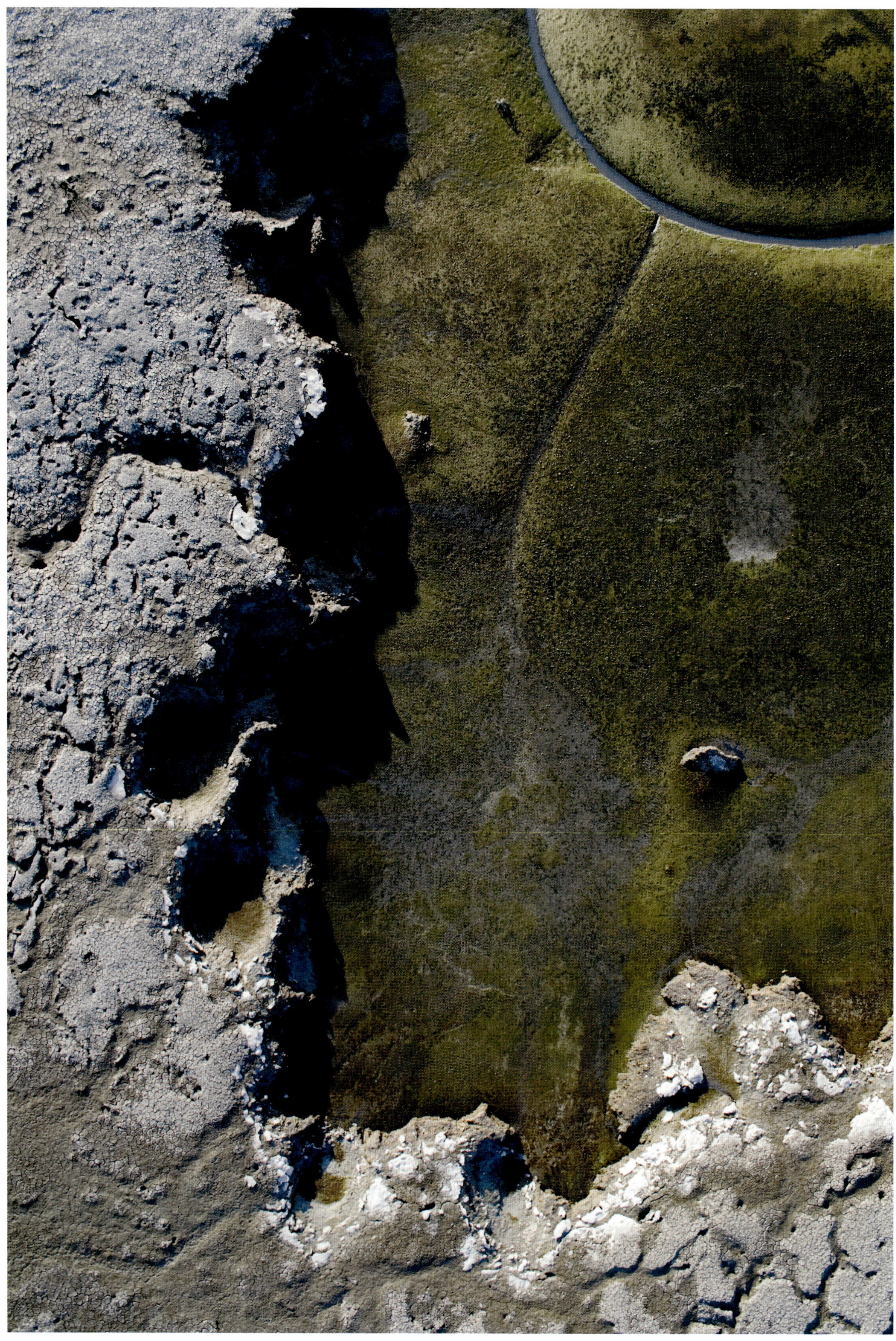

Plate 70. Fresh dredge marks (Salt Pond A21, April 2008)

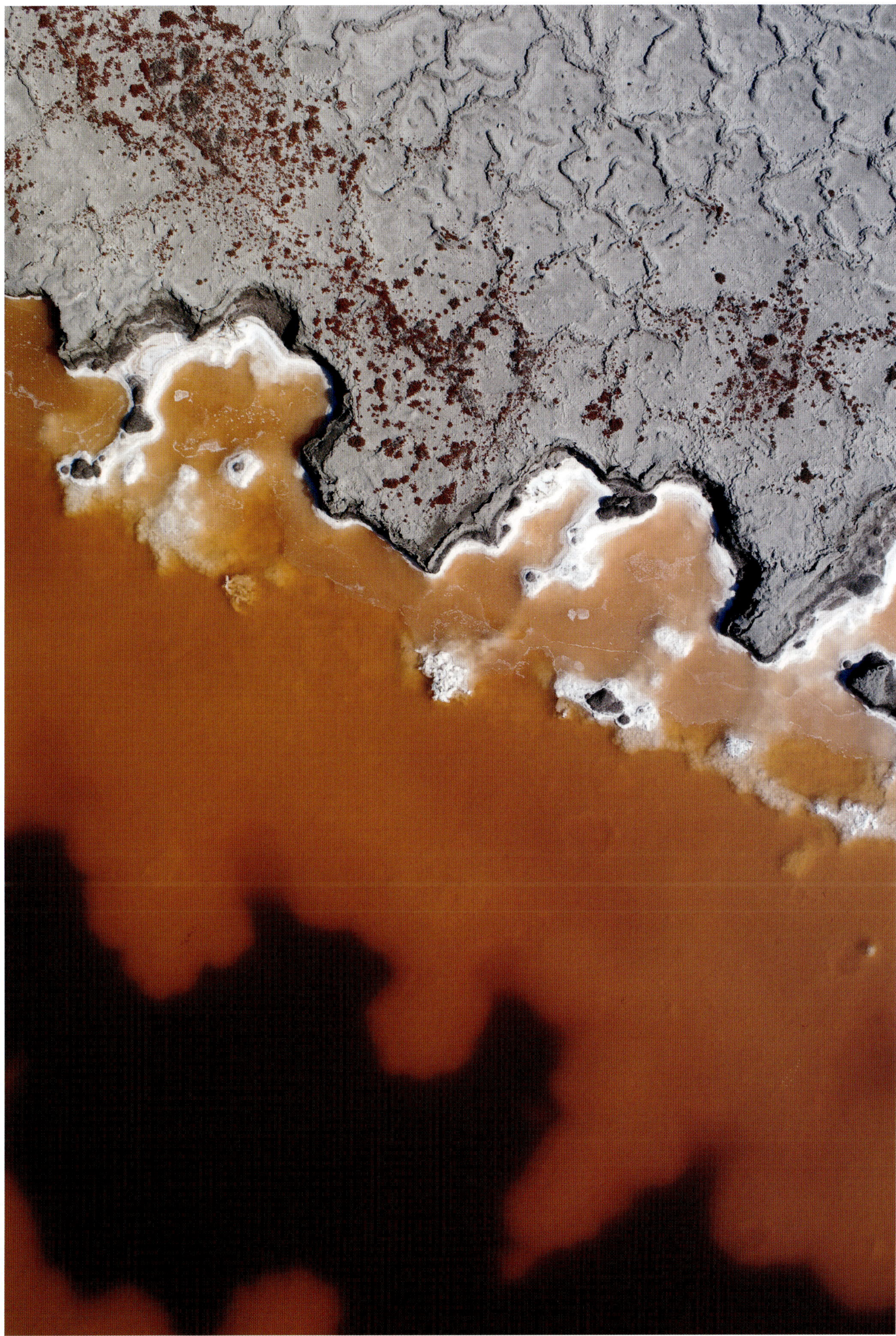

Plate 71. Two tiers of dredge marks (Salt Pond R5, July 2009)

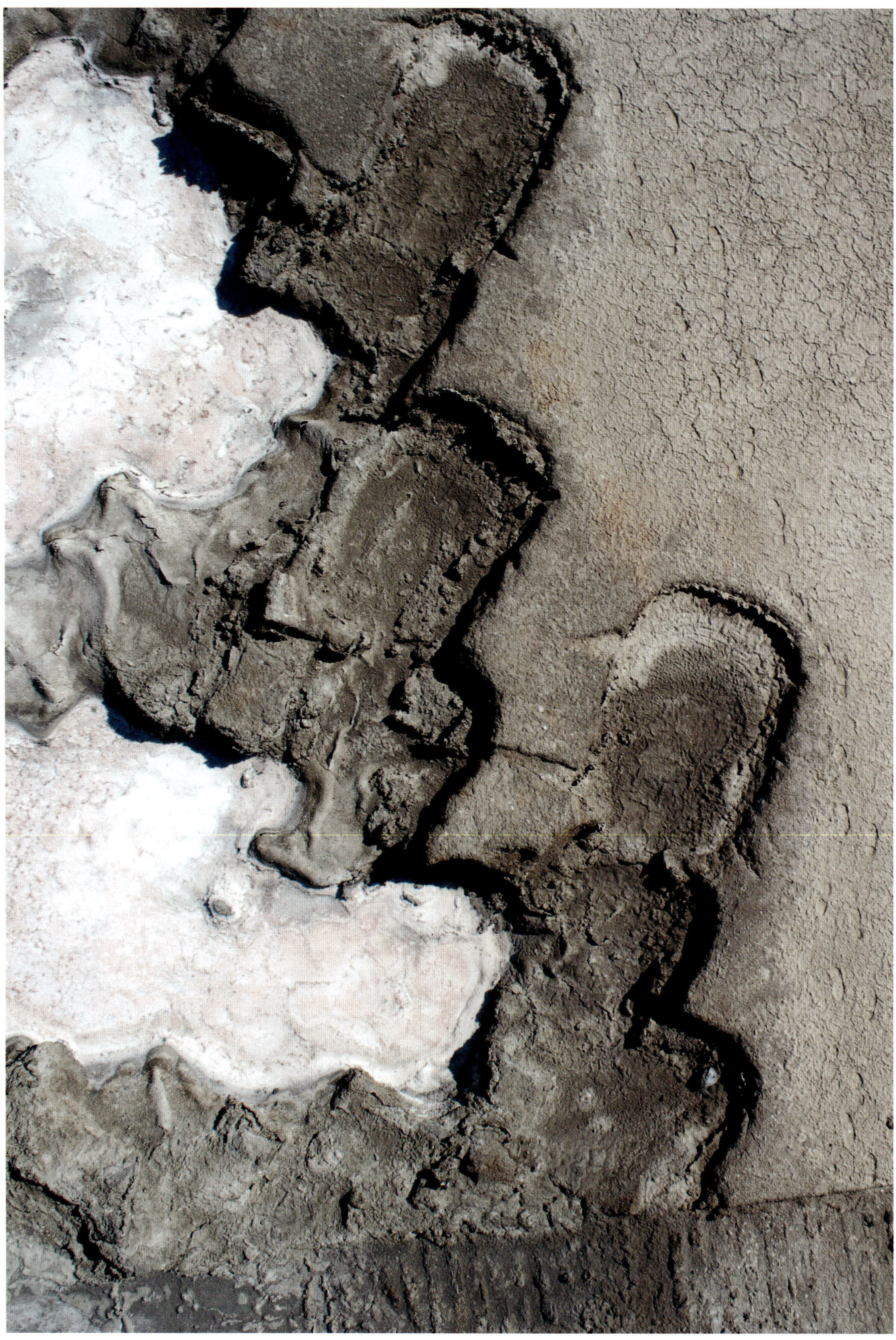

Plate 72. Sharp marks in dry pond bottom (Salt Pond SF2, August 2009)

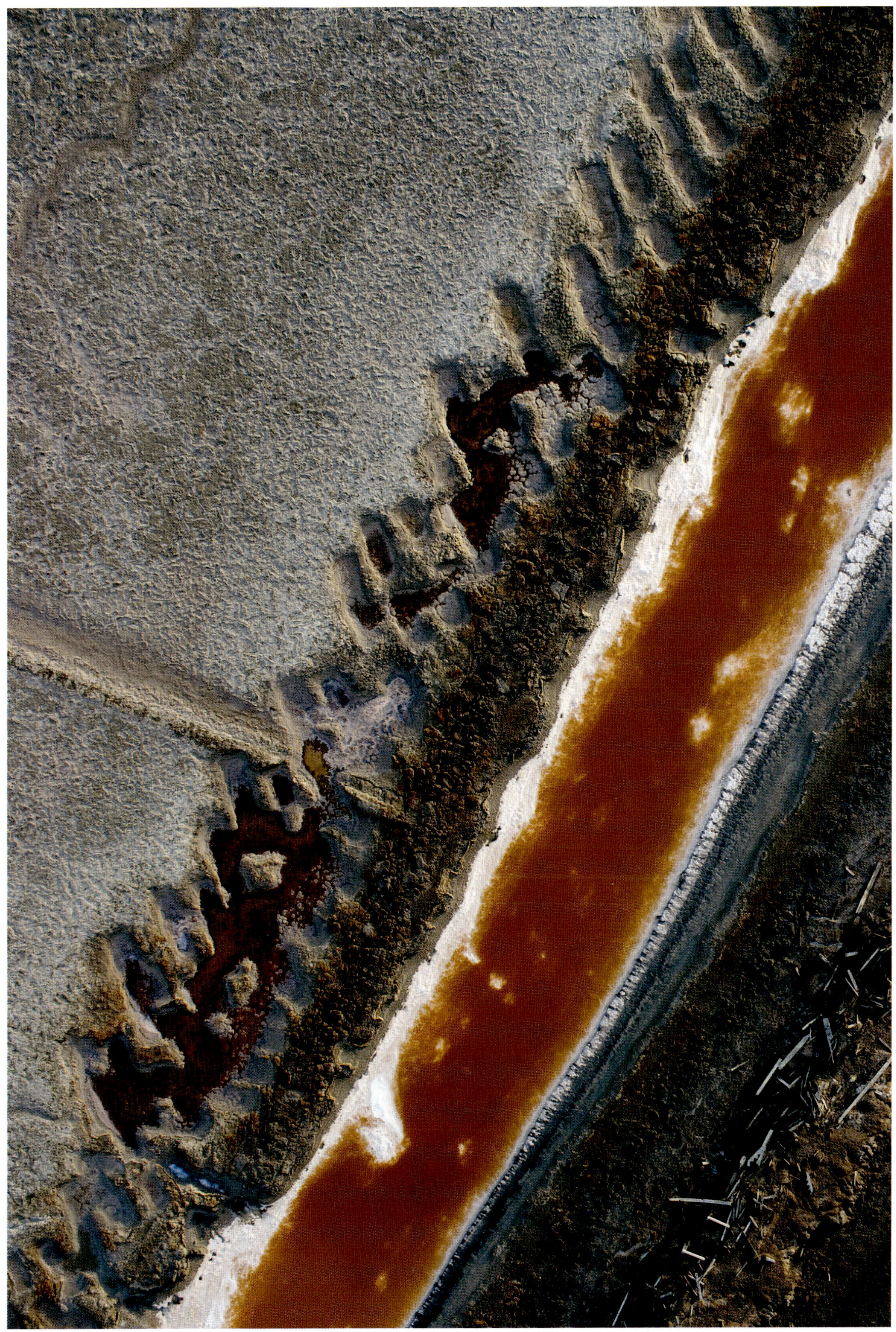

Plate 73. High view of dredge marks (Salt Pond E14, September 2009)

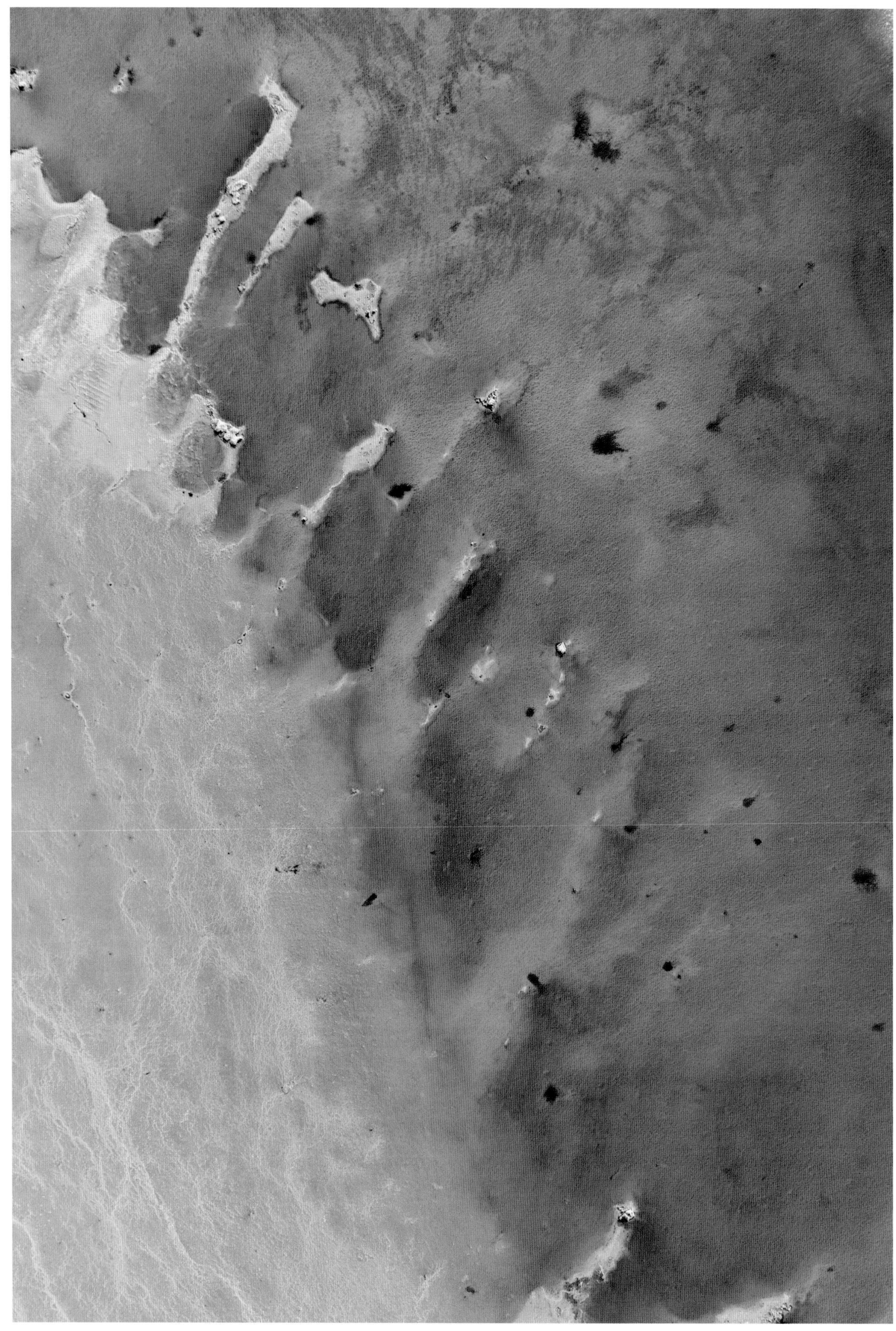

Plate 74. Vague marks revealing age (Salt Pond A23, October 2011)

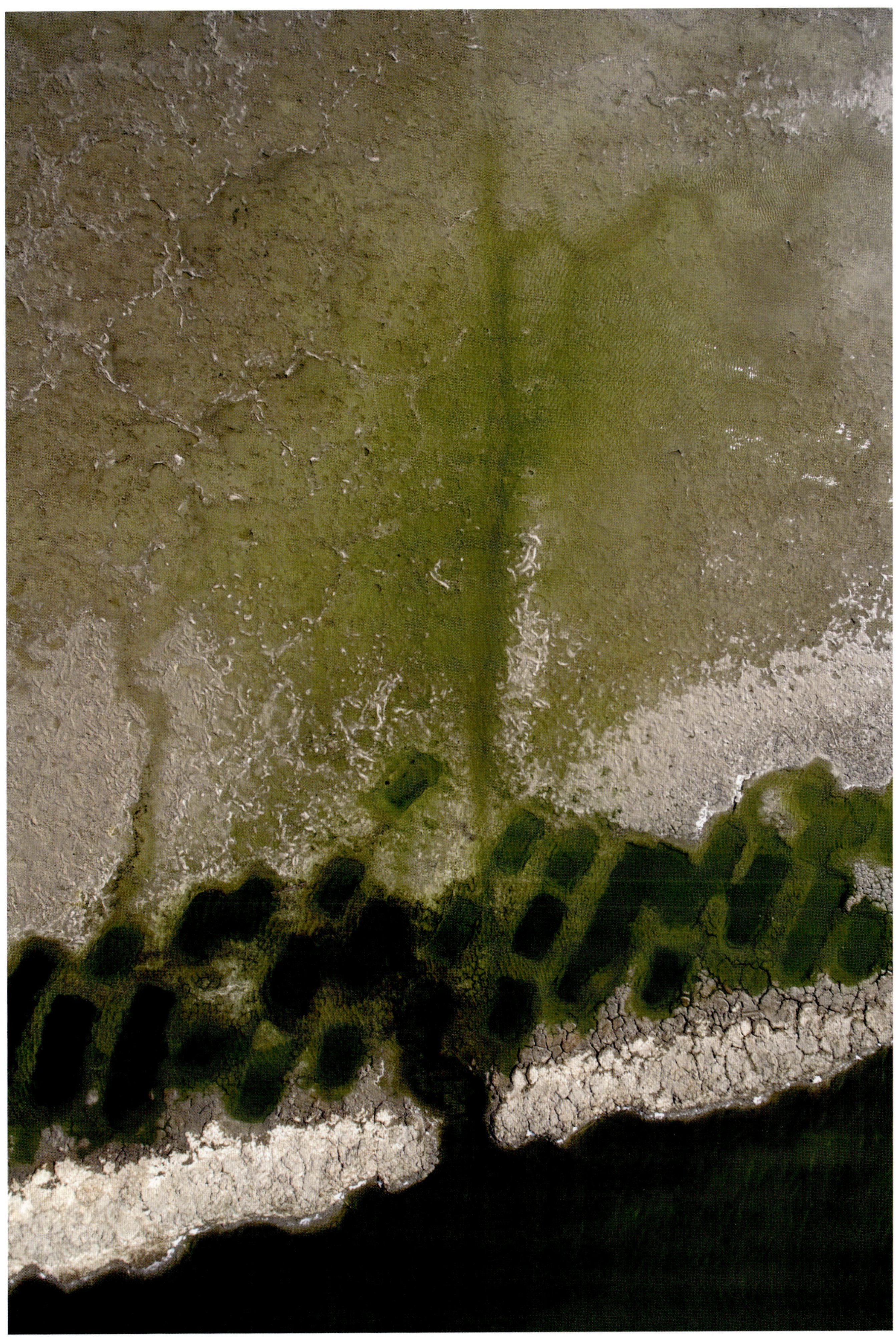

Plate 75. Marks in a secondary borrow ditch (Salt Pond E14, September 2008)

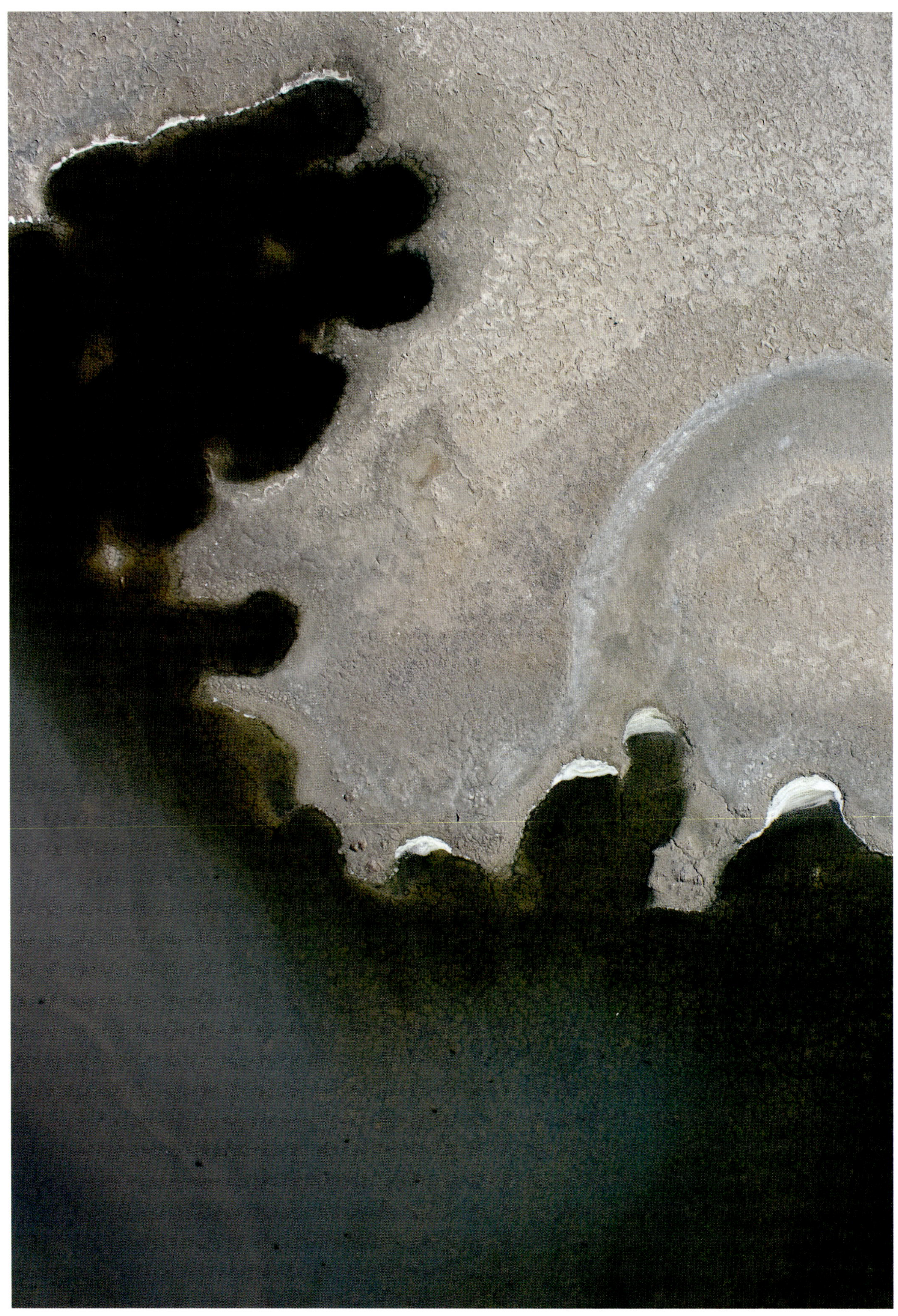

Plate 76. Dredge marks bunched near levee bend (Salt Pond E14, October 2008)

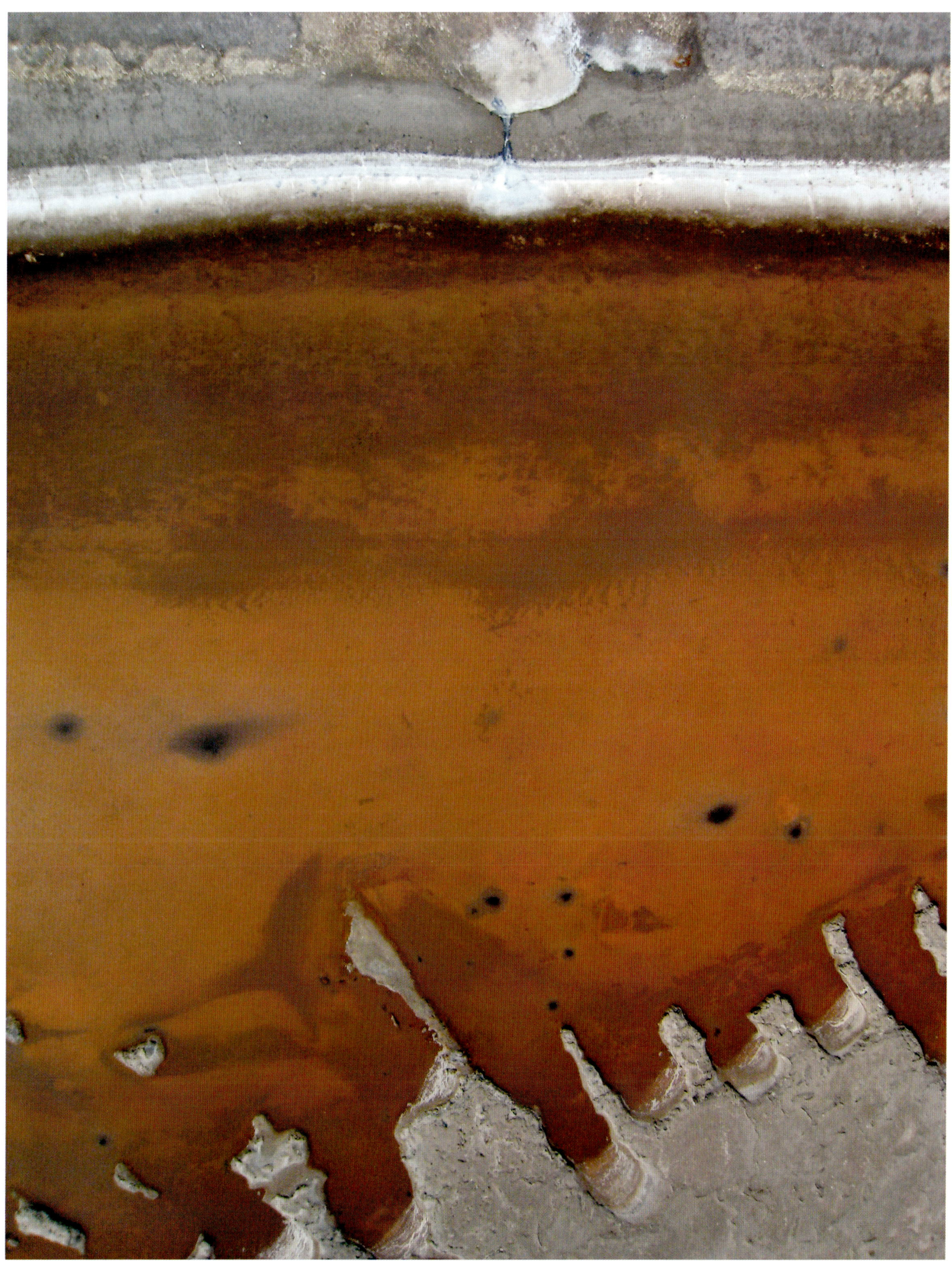

Plate 77. Sharp marks along a shallow ditch (Salt Pond A23, April 2007)

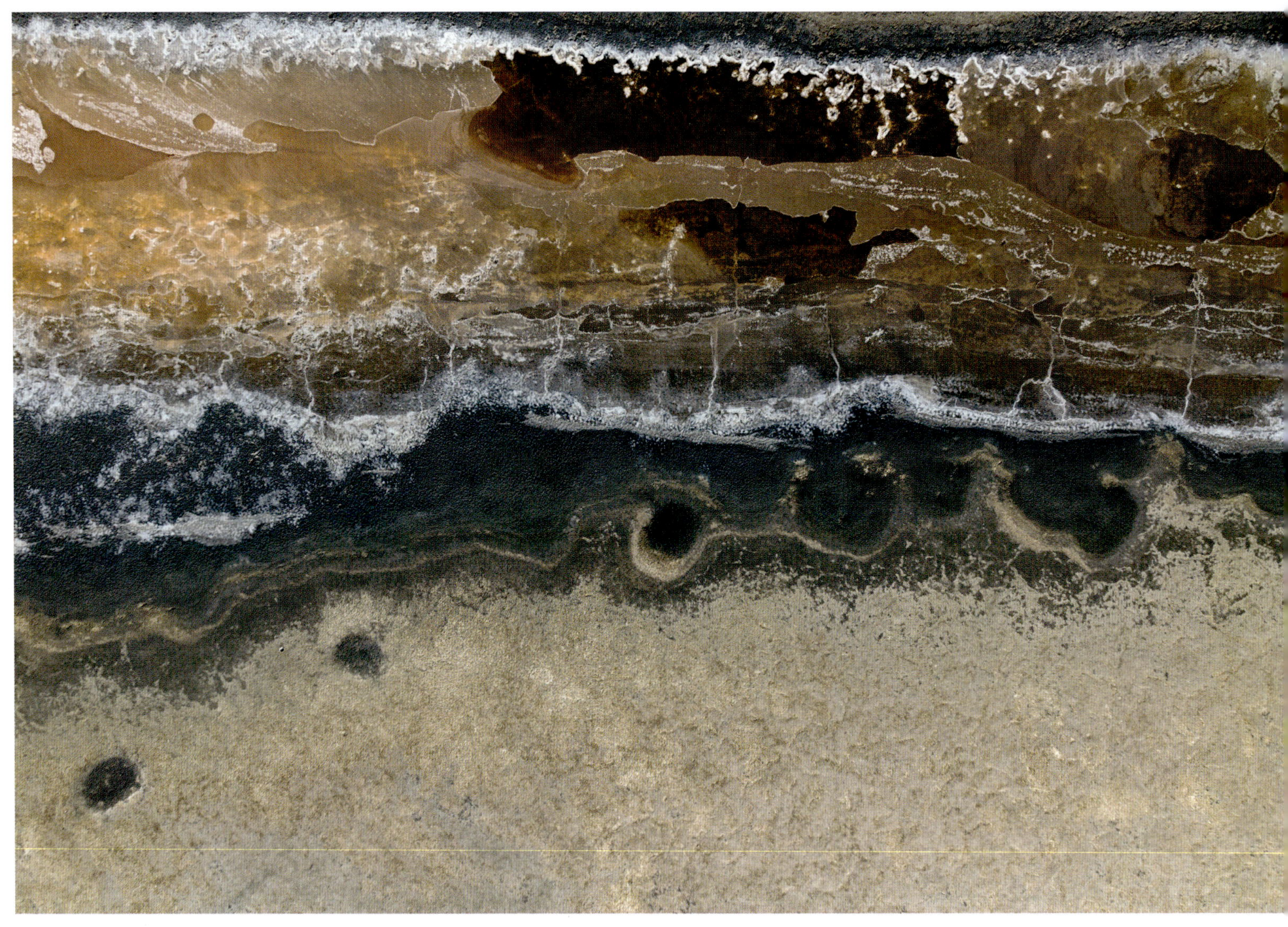

Plate 78. Dredge marks have almost disappeared (Salt Pond E14, September 2010)

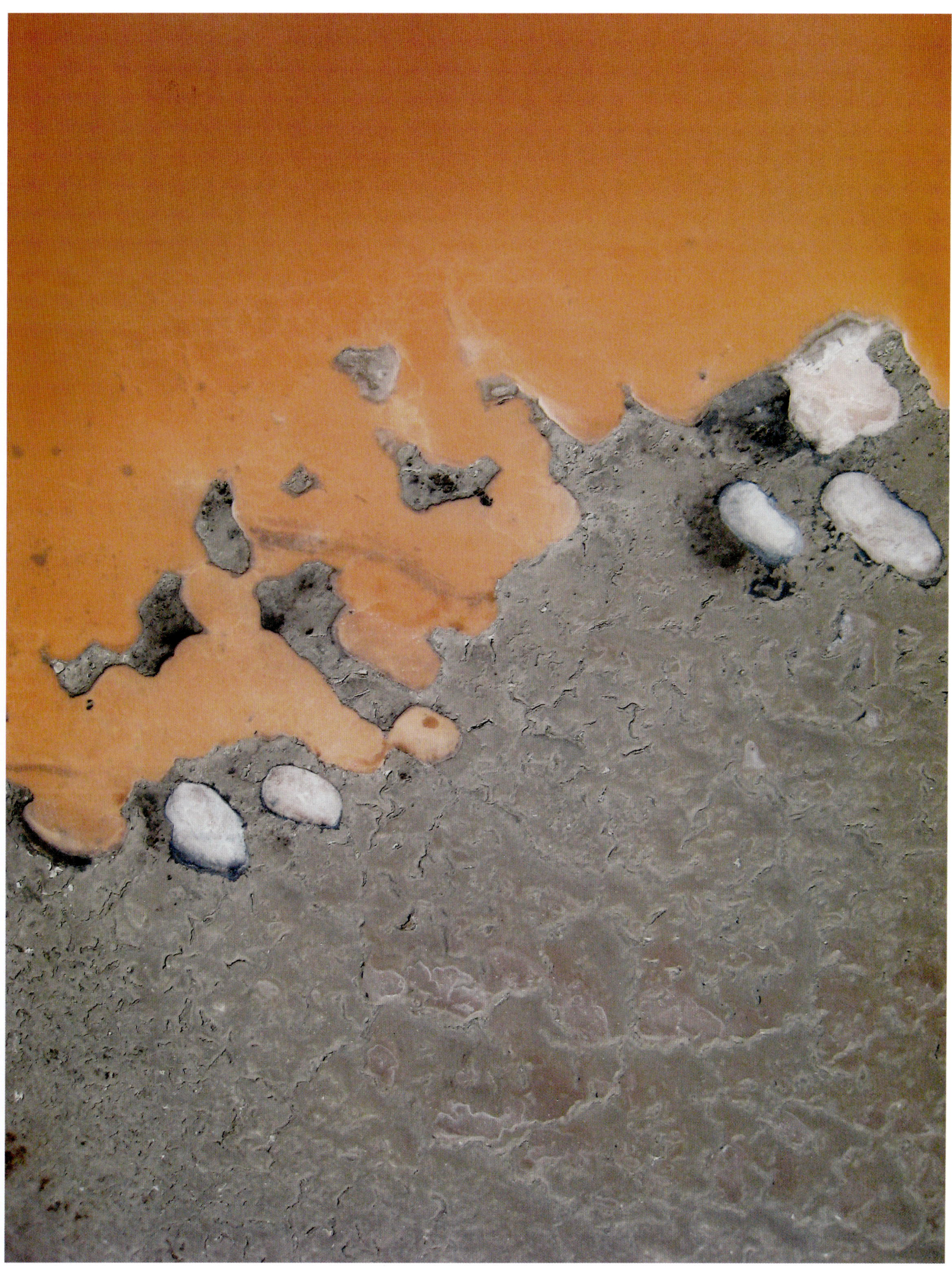

Plate 79. Soft marks suggesting age (Salt Pond A23, May 2007)

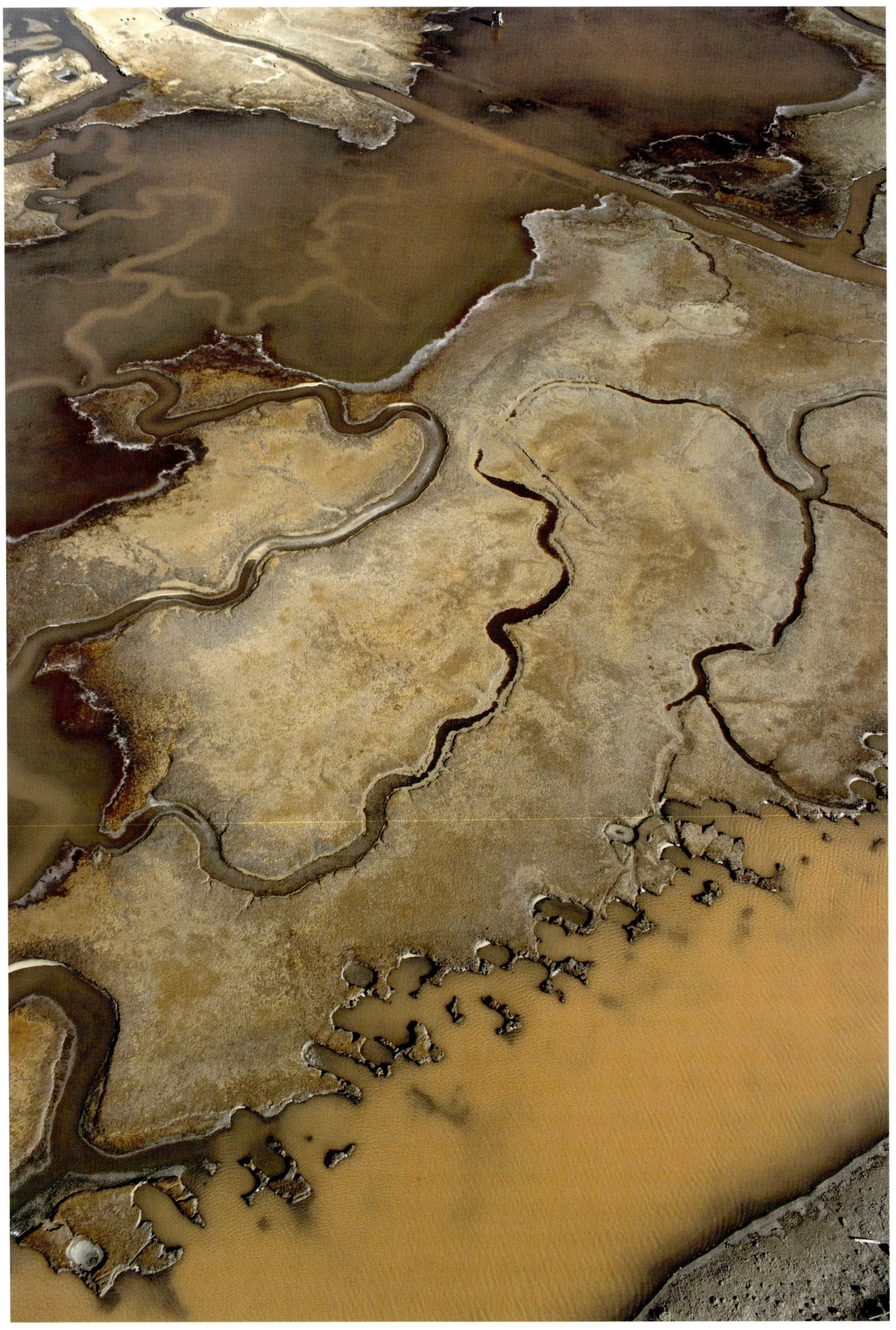

Plate 80. Dredge marks provide height clue (Salt Pond E8, October 2009)

California has long served as the historical stopover and wintering ground for waterfowl making their way from the cold wilds of Alaska, Canada, and Siberia: a migration that includes vast numbers of ducks. Migration patterns established over several millennia changed with the arrival of Europeans. By the late nineteenth century, the large wetlands that served the great Pacific Flyway were disappearing from the map at a rapid pace. Areas like the Klamath, Buena Vista, and Tulare lake regions were reclaimed for agriculture, as were large tracts in the Sacramento–San Joaquin Delta. As habitat dwindled, the marshlands near Eden Landing were considered the best of all locations for waterfowl, and thus the hunting of ducks. Duck hunting continued in full force even as these marshes were being diked for salt production. In fact, the levees and salt ponds created by the early saltworks offered access to prime hunting grounds and duck hunting came to coexist, sometimes awkwardly, side by side with the saltworks.

As I come across ducks on my levee walks, they strike me as a particularly wary species prone to flushing like a covey of huge startled quail, and perhaps with good reason. In the late nineteenth century market hunters harvested ducks with large-bore shotguns, duck blinds offered effective cover for a multitude of recreational hunters, and ducks were shot in annual numbers that literally

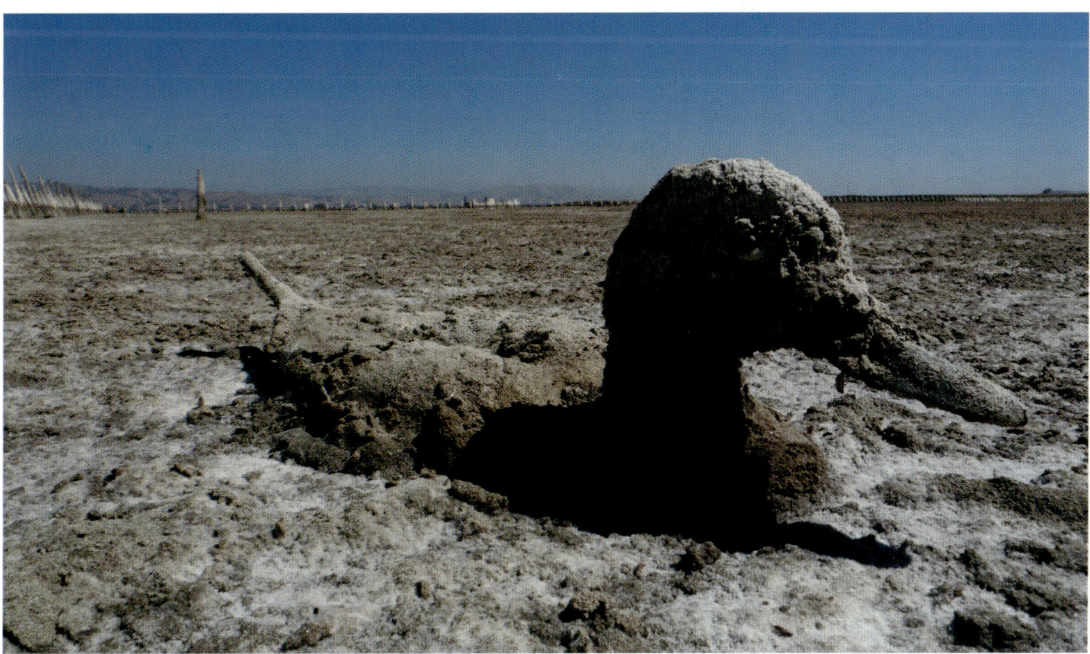

A traitorous decoy: one frequently finds abandoned duck decoys in the South Bay landscape, testimony to a tradition of duck hunting that continues today.

exceeded a quarter-million in the San Francisco Bay Area alone. Public concern over the decline of once vast waterfowl populations due to loss of habitat and aggressive commercial hunting led to 1915 legislation banning the commercial sale of ducks, placing an increased emphasis on scientific investigation in the management of wildlife, and ultimately establishing wildlife refuges designed to provide habitat for waterfowl. By 1950, refuges and waterfowl management areas encompassed most of the California habitat still available to the Pacific Flyway. Ducks are hunted to this day in these publicly managed lands, including sections of the South Bay, under policies designed to be sustainable. Quacker vigilance is still very much warranted.

In reviewing my South Bay aerials I find, like drawings on another layer of tracing paper, an ordered schema related to the hunting of ducks. The current-day blinds, crude shacks on stilts, are easily evident from the ground and are spaced about the landscape at shotgun-range intervals. Much less evident to the casual observer is a network of small blinds that are essentially holes, lined with wood slats or occasionally concrete, in the bottom of salt ponds or marshland. These dugout blinds are often situated in a mound constructed to raise the blind's opening above water level. I first encountered this construction in the bottom of fallow salt ponds near Warm Springs and remained puzzled for over a year by the curious array of what looked like twin wooden barrels sunk into hummocks. The hummocks were clearly constructed features, as evidenced by a star-burst pattern of scrape marks surrounding each mound, although they were oddly adjacent to the similar natural mounds of the Warm Springs vernal pool landscape. I eventually approached one of these sites to find the remains of an old chair and shelf in the wooden barrel. Once aware of this form of duck blind, I began to identify others in many of my pho-tographs and appreciate their broad distribution.

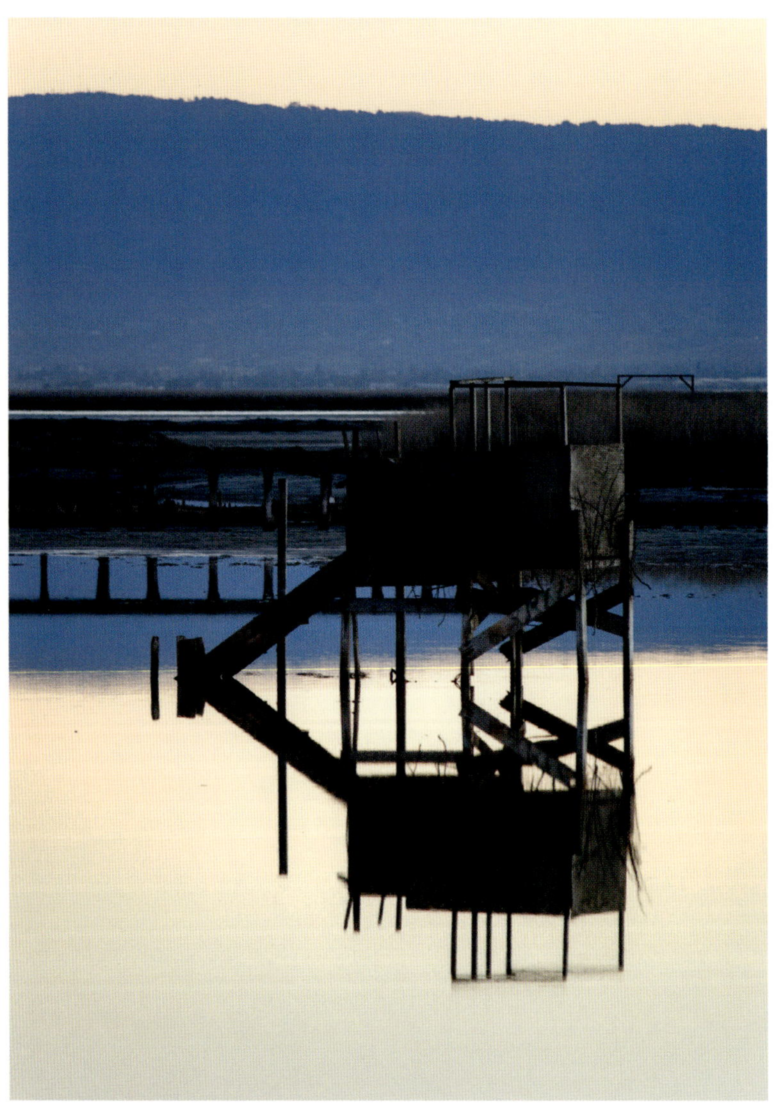

Ground-level view of a contemporary, raised duck blind.

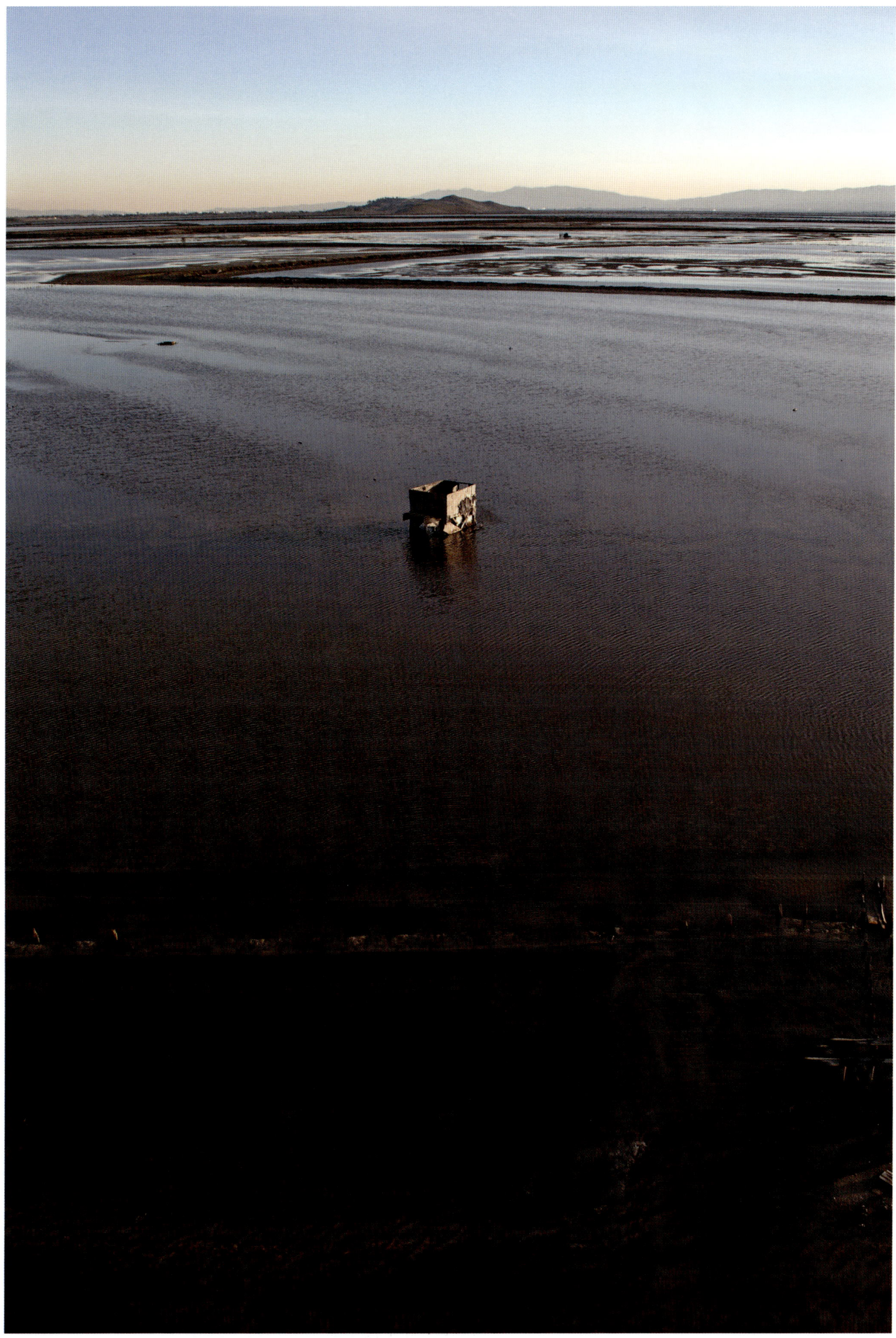

Plate 81. Raised blind in a flooded pond (Salt Pond E14, January 2009)

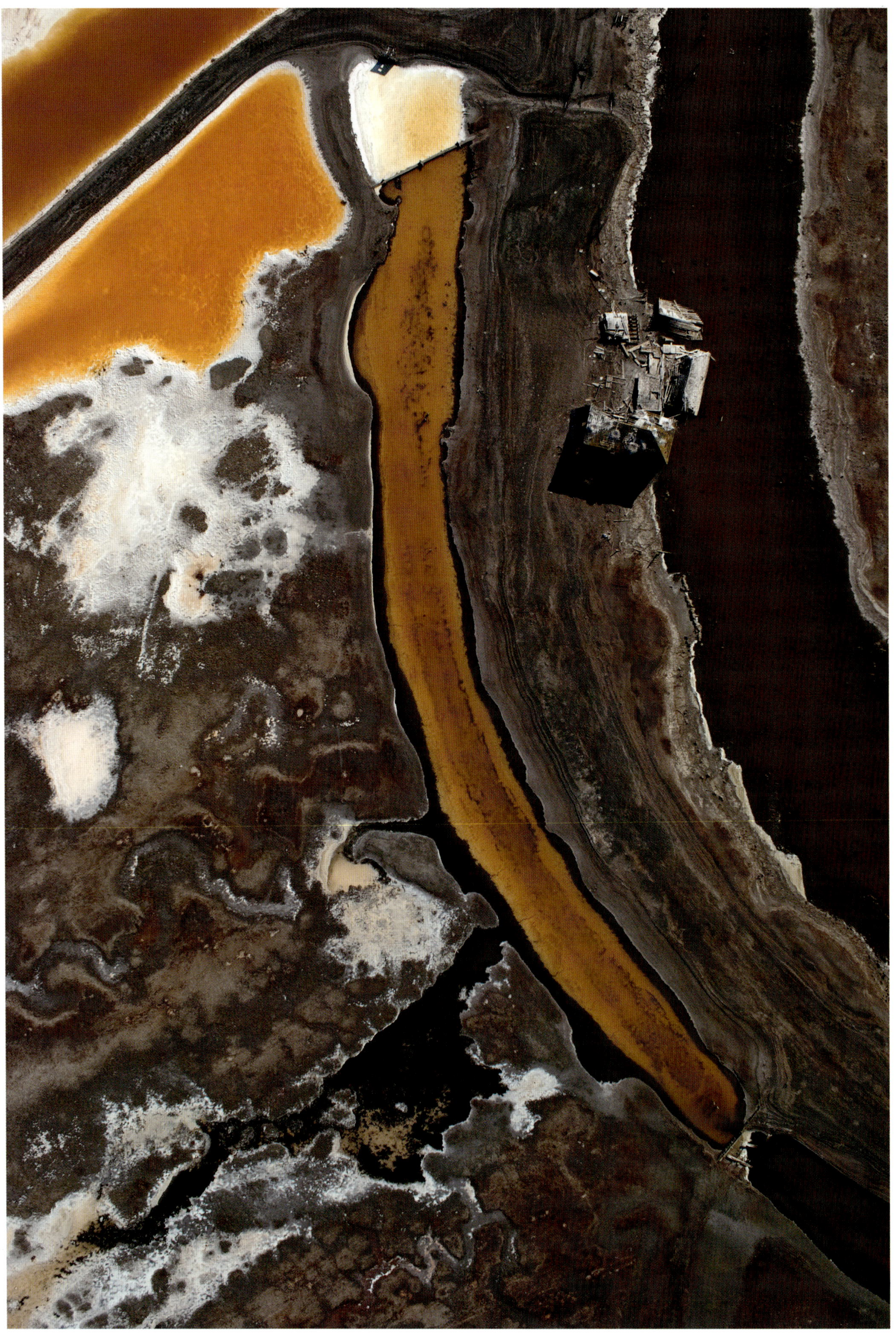

Plate 82. Duck hunter's cabin in ruins (Salt Pond E6B, September 2009)

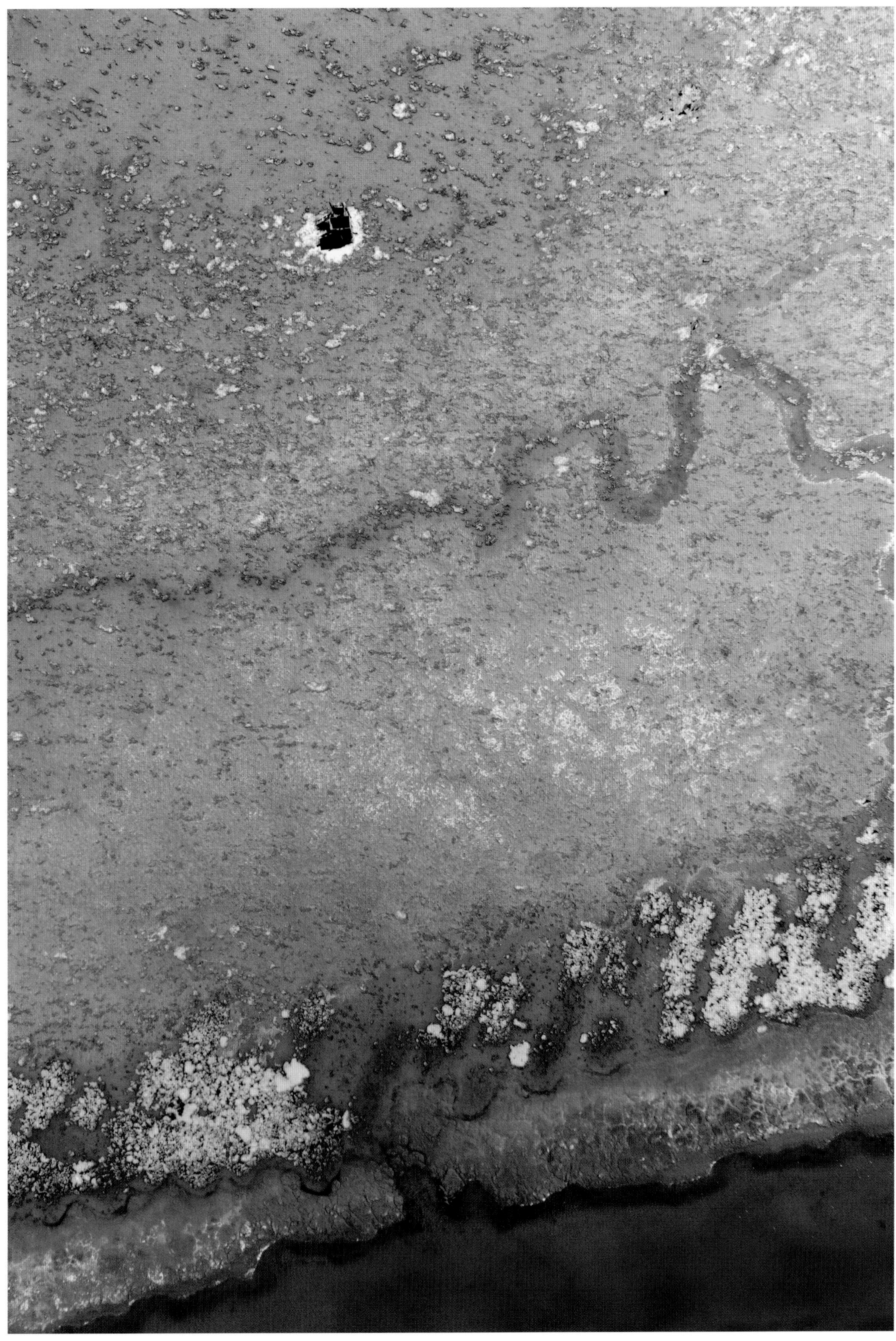

Plate 83. A small raised duck blind sited near a row of dredge marks (Salt Pond E14, August 2011)

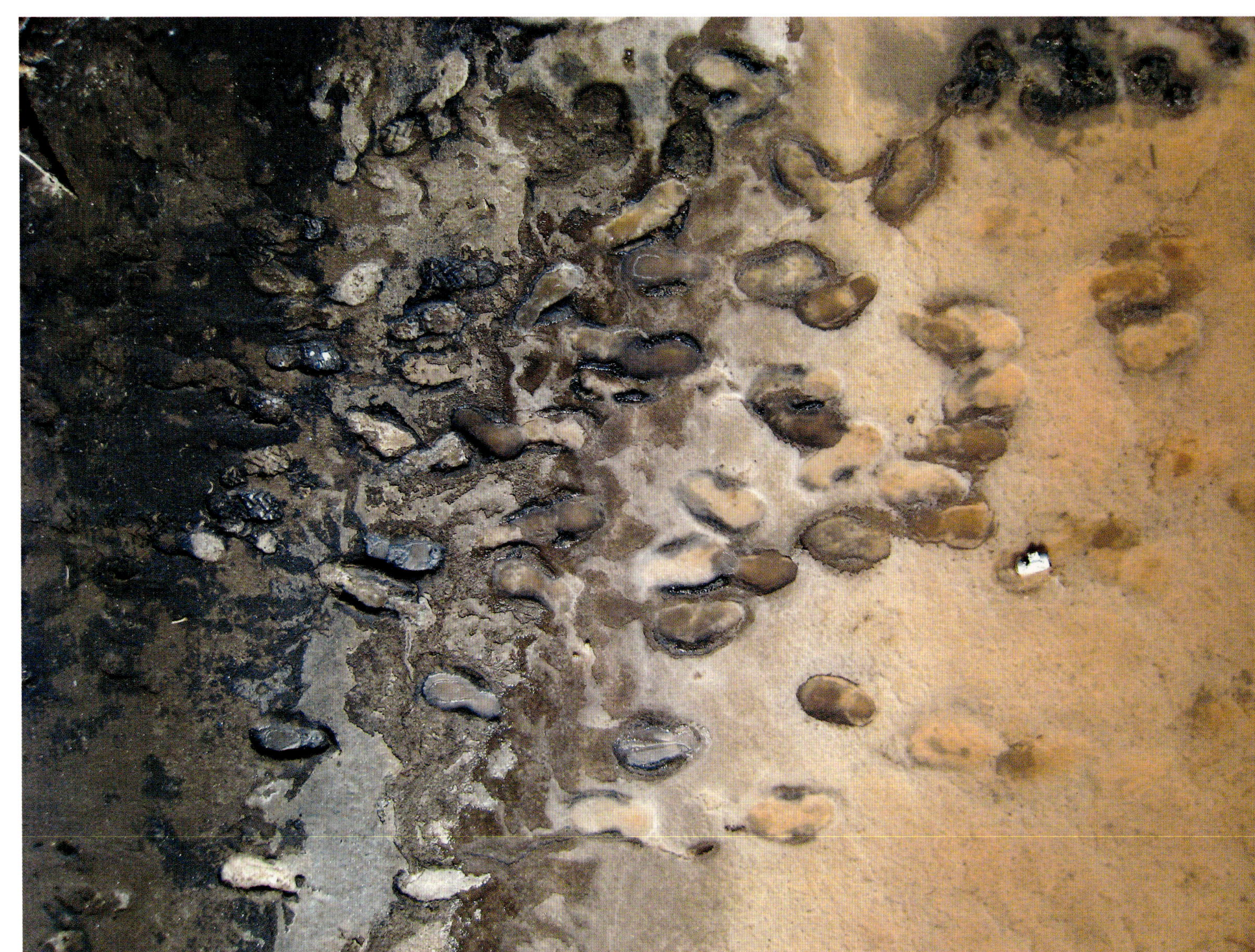

Plate 84. Duck hunter tracks (Salt Pond A23, May 2007)

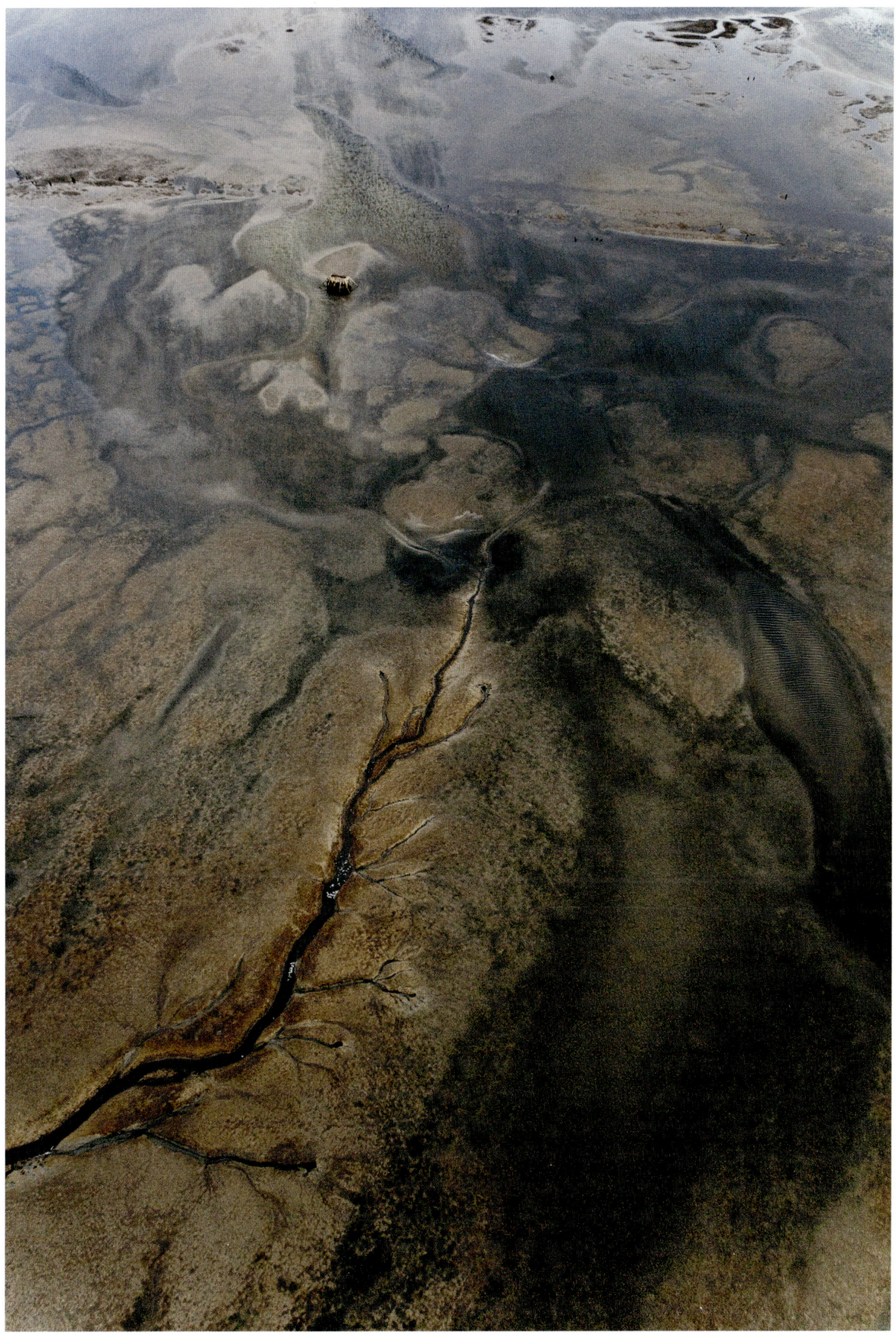

Plate 85. Raised blind at low tide (Mount Eden Creek mudflat, October 2009)

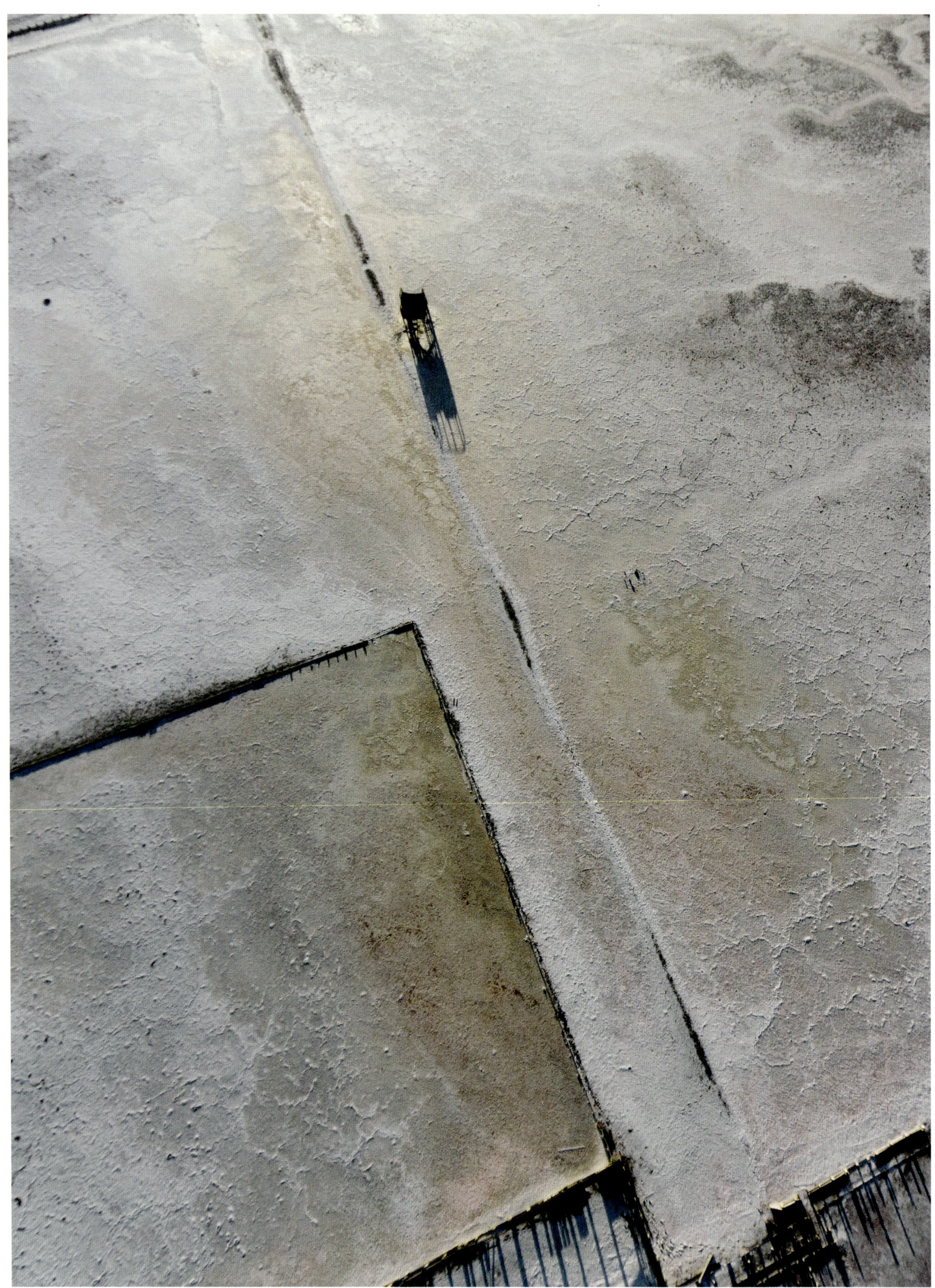

Plate 86. Contemporary blind on stilts (Salt Pond E13, September 2010)

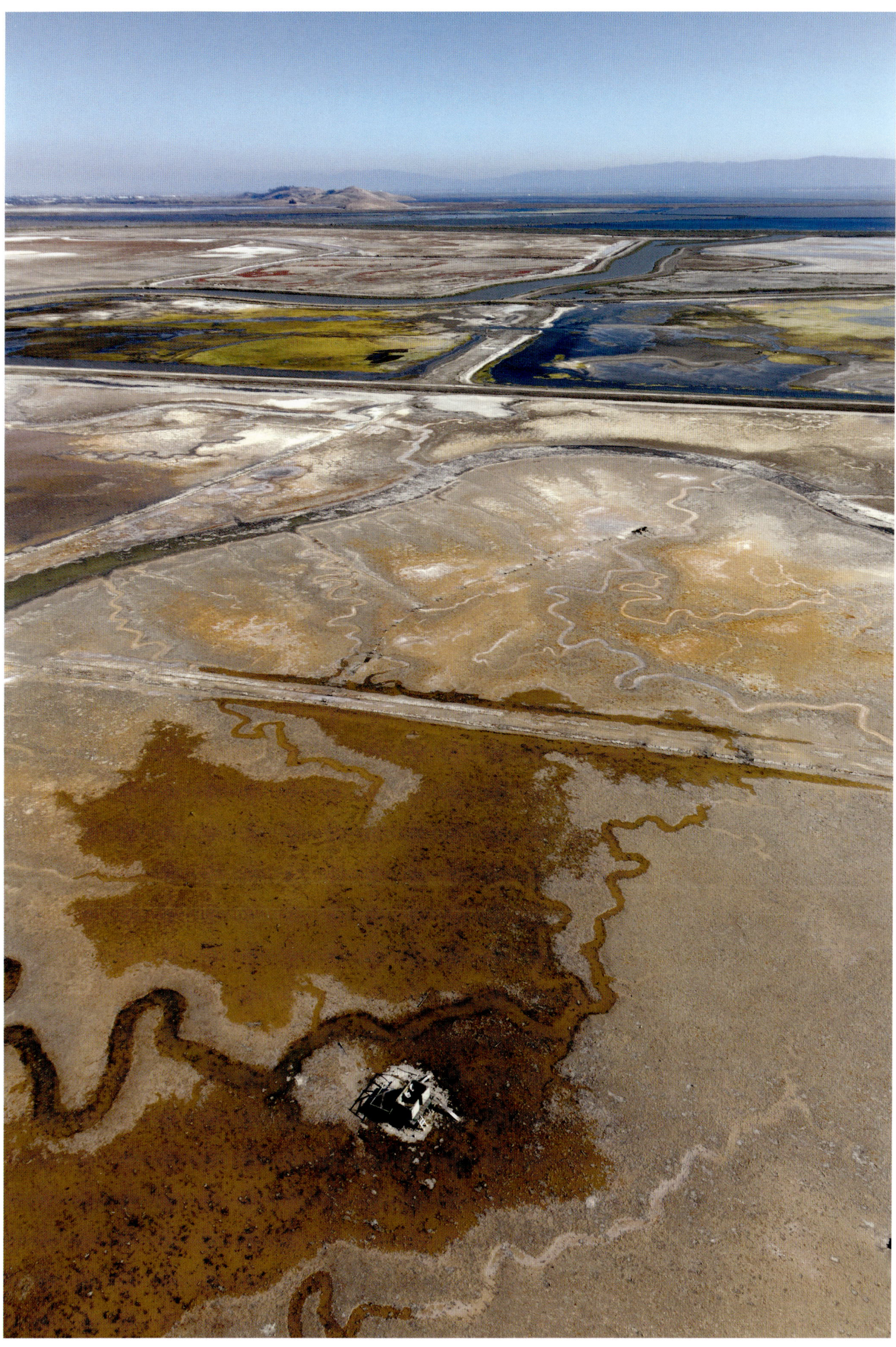

Plate 87. Small duck blinds dot the landscape (Salt Pond E14, August 2011)

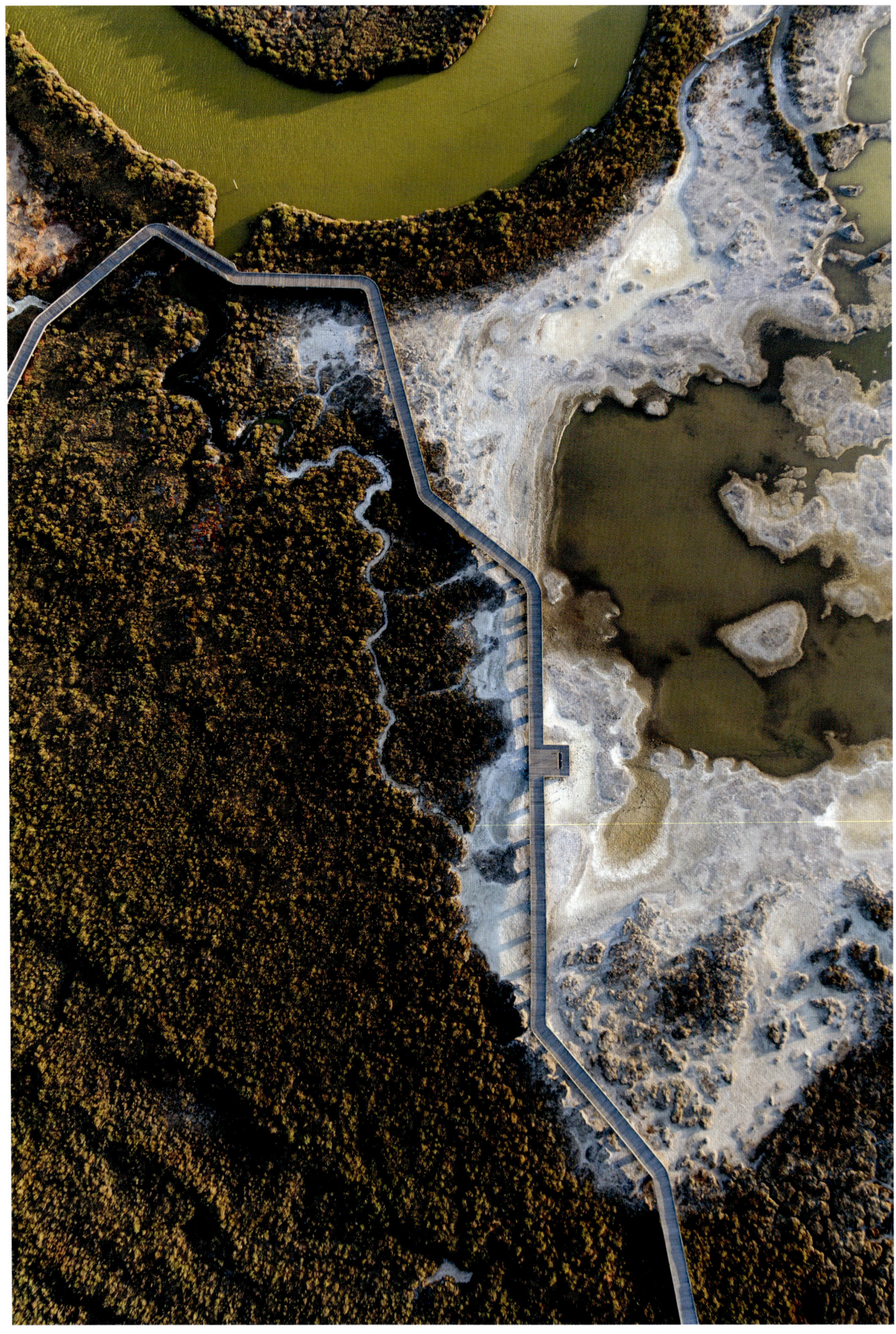

Plate 88. Environmental Education Center boardwalk (New Chicago Marsh, September 2011)

6: OUR DAY

HAROLD GILLIAM'S 1957 book *San Francisco Bay* offers an affectionate portrait of the bay at a pivotal moment in its history. By this date the salt industry had been consolidated for two decades under Leslie Salt, whose enterprise carpeted most of the South Bay landscape with large evaporation ponds. In the aftermath of the Depression and World War II there was, it seemed, a widespread societal hunger for modern progress. Further commercial development of the bay lands seemed inevitable and all indications suggested it would be bold. For instance, Gilliam describes John Reber's plan to convert the South Bay and San Pablo Bay into freshwater lakes through the creation of unimaginably large dams, an idea that prompted the Corps of Engineers' giant analog Bay Model in Sausalito. Yet Gilliam presents these plans and the likely trajectory of progress without enthusiasm, reserving his superlatives for descriptions of the bay's beauty and historic nature, qualities that seemed very much in peril. He was not alone in these concerns.

Three years after publication of his book, Harold Gilliam participated in a seminal meeting organized by soon-to-be activists Sylvia McLaughlin, Kay Kerr, and Esther Gulick. This storied gathering of conservationists marked the beginning of the Save the Bay movement, itself the beginning of a larger awakening of environmentalists. Against incredibly long odds, Save the Bay marshaled skill, persuasion, and determination to advocate for environmental policies limiting wetland development. They succeeded in establishing policies for development along the Berkeley waterfront and then throughout the entire Bay Area. The grassroots movement expanded far beyond a coalition of birders, sportsmen, and conservationists by appealing directly to the public at large with cogent arguments addressing the environmental consequences of unbridled development. They were wildly successful. Within fifteen years legislation had established the unprecedented Bay Conservation and Development Commission to regulate development adjacent to San Francisco Bay; enacted clean water acts at the state and federal levels; and created the San Francisco Bay National Wildlife Refuge. If the mantra of the 1950s had been "you can't stop progress," the environmental movement's genius lay in providing, with care and consideration, an alternative vision of progress.

By 1980 the San Francisco Bay National Wildlife Refuge had acquired title to over half of the South Bay salt ponds, acquisitions in which Cargill Salt retained perpetual operating rights to work the ponds for salt production. In essence the refuge served much like an open space land trust, removing the specter of further development without changing the everyday use of the ponds. Around this time a series of relatively small marsh restoration projects, often associated with environmental mitigation settlements, offered the first evidence, qualified with many cautions, that marshes lost to diking could be reestablished. From these revelations built a slow but firm conviction that large areas of the South Bay could be restored. After the due diligence of feasibility studies and goal setting, an ambitious deal was struck in 2003 between Cargill Salt and a coalition of state, federal, and local governments aided by private foundations. The refuge, or more properly the coalition under the rubric of the South Bay Salt Pond Restoration Project, was now in the business of wetland restoration at a massive scale and the South Bay landscape would soon become much more complex. While the new order would include remnants of original marsh and a network of active salt ponds covering about a third of their former extent, well over half of the former South Bay wetlands were now open to redefinition.

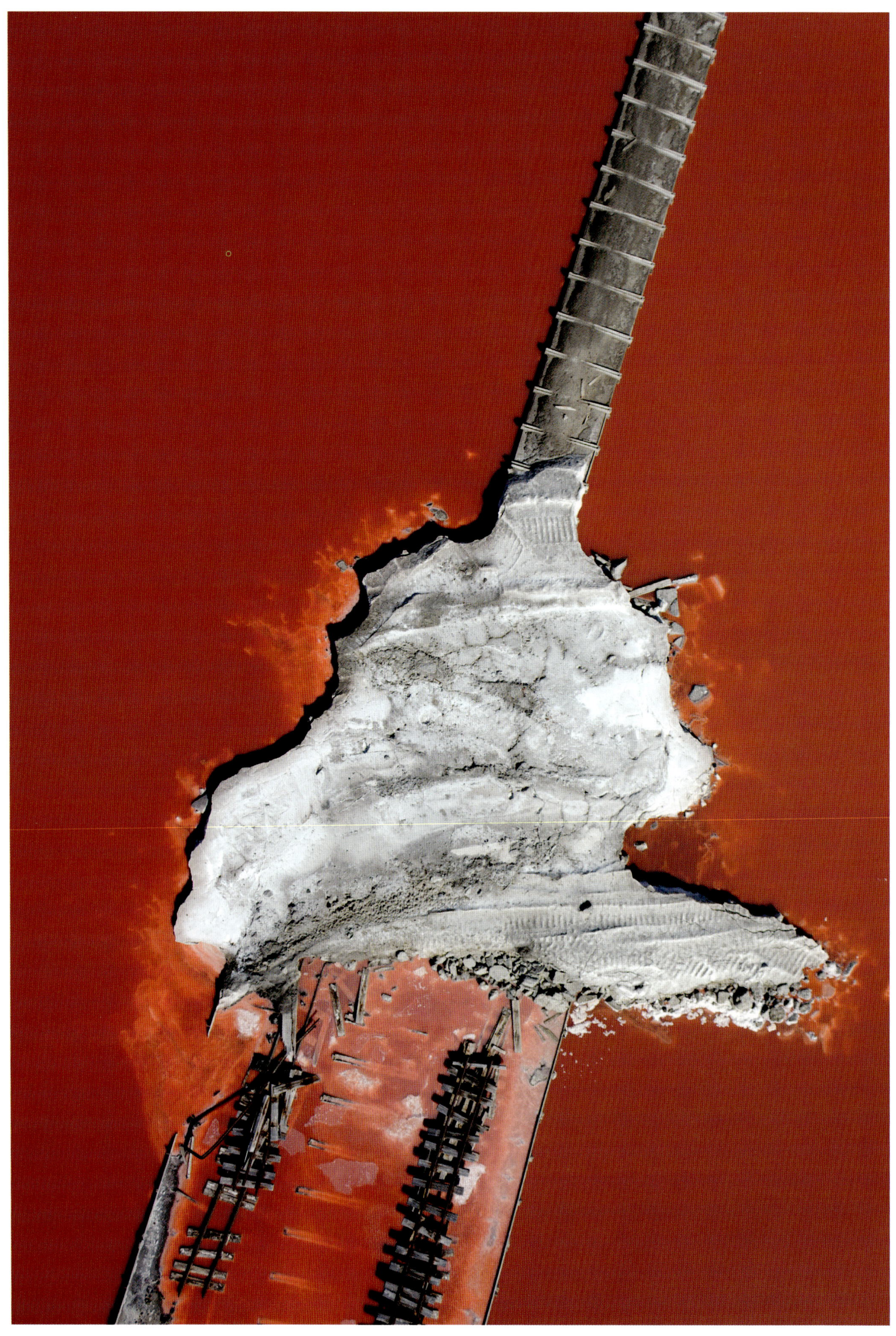

Plate 89. Dike dividing crystallizer beds (Newark Crystallizer Beds, August 2010)

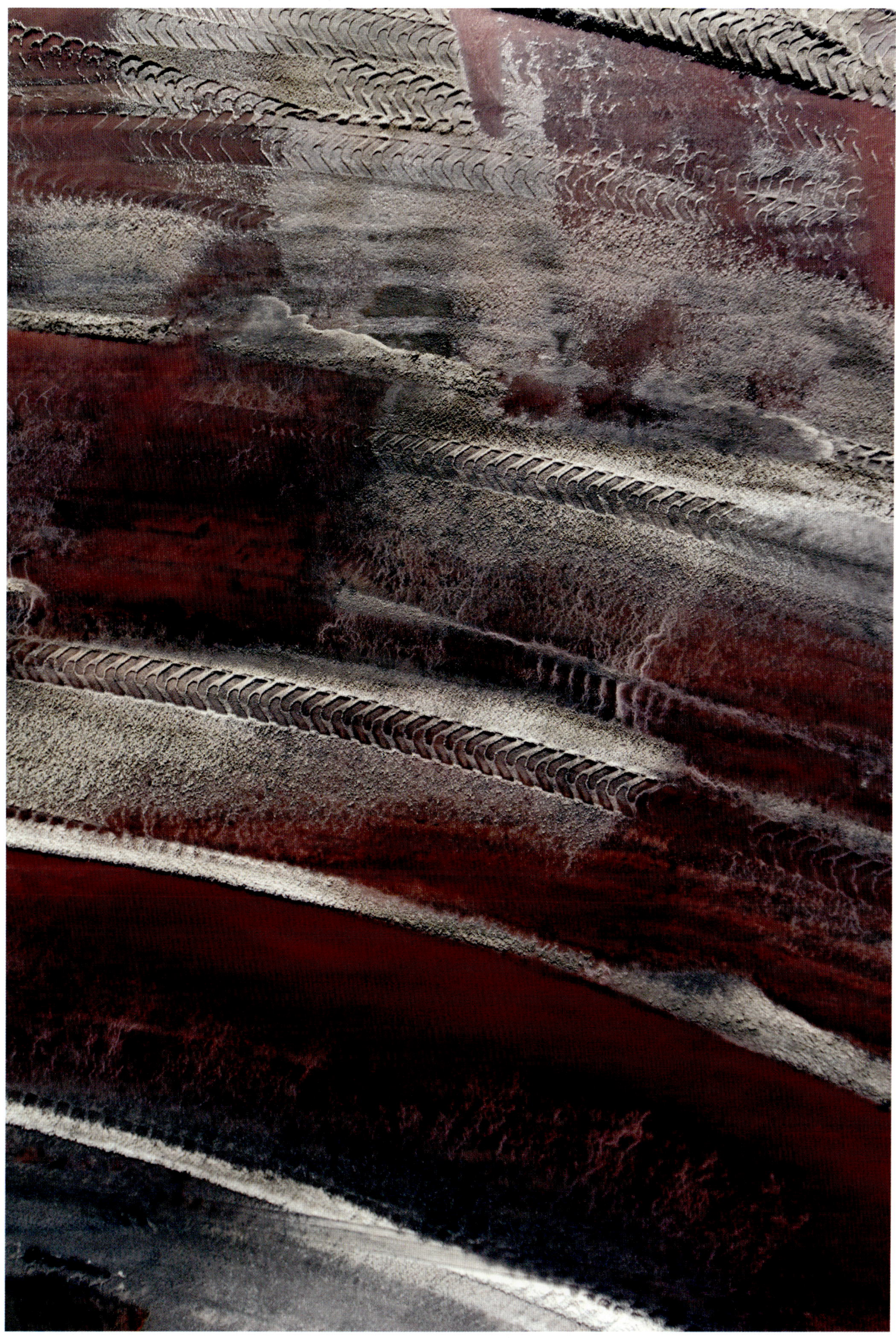

Plate 90. Surface texture of harvested bed (Newark Crystallizer Beds, November 2010)

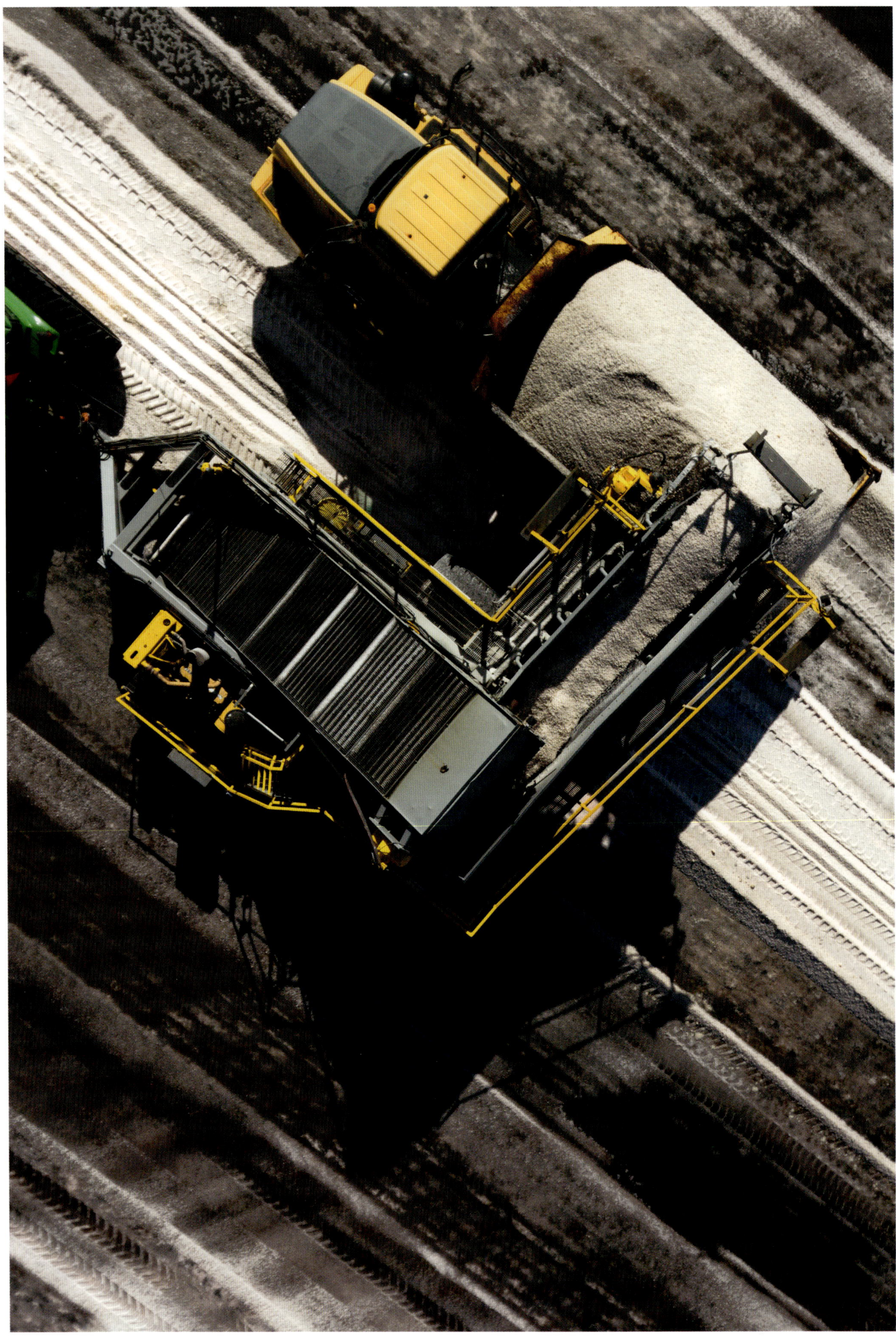

Plate 91. Harvester loading a dump truck (Newark Crystallizer Beds, October 2010)

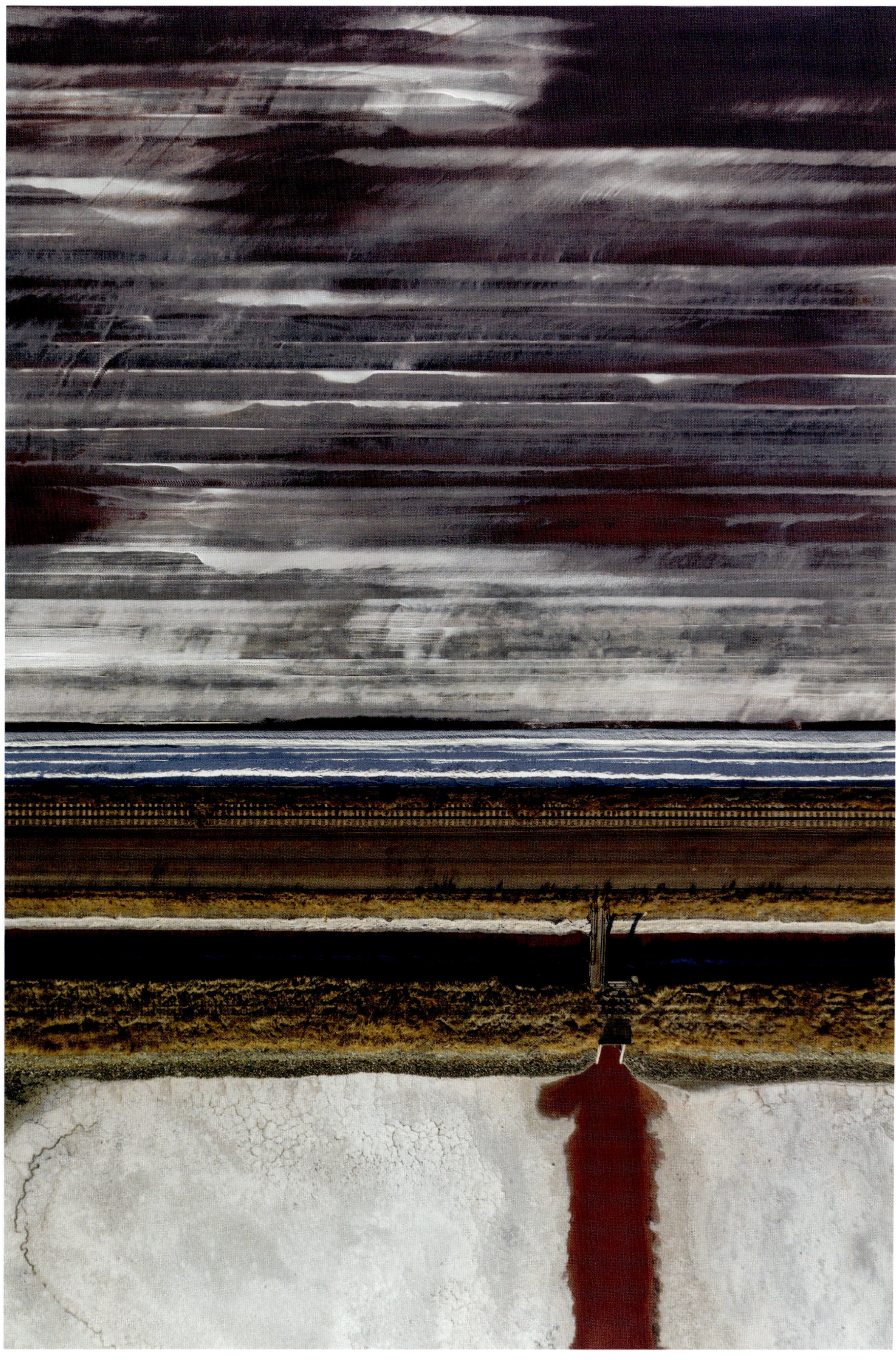

Plate 92. A compact and orderly layout supports the salt harvest (Newark Crystallizer Beds, November 2010)

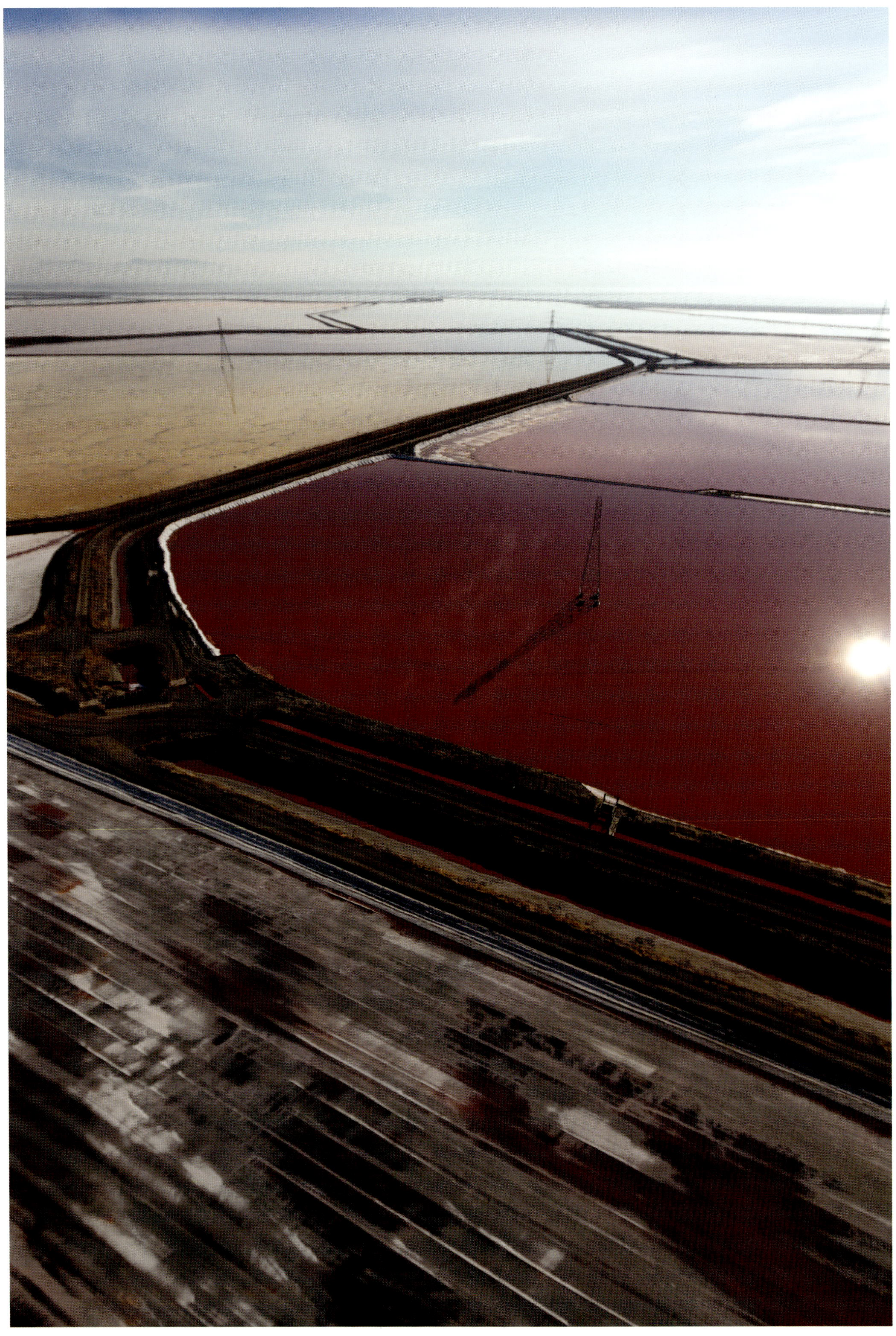

Plate 93. A carefully orchestrated landscape (Newark Crystallizer Beds, November 2010)

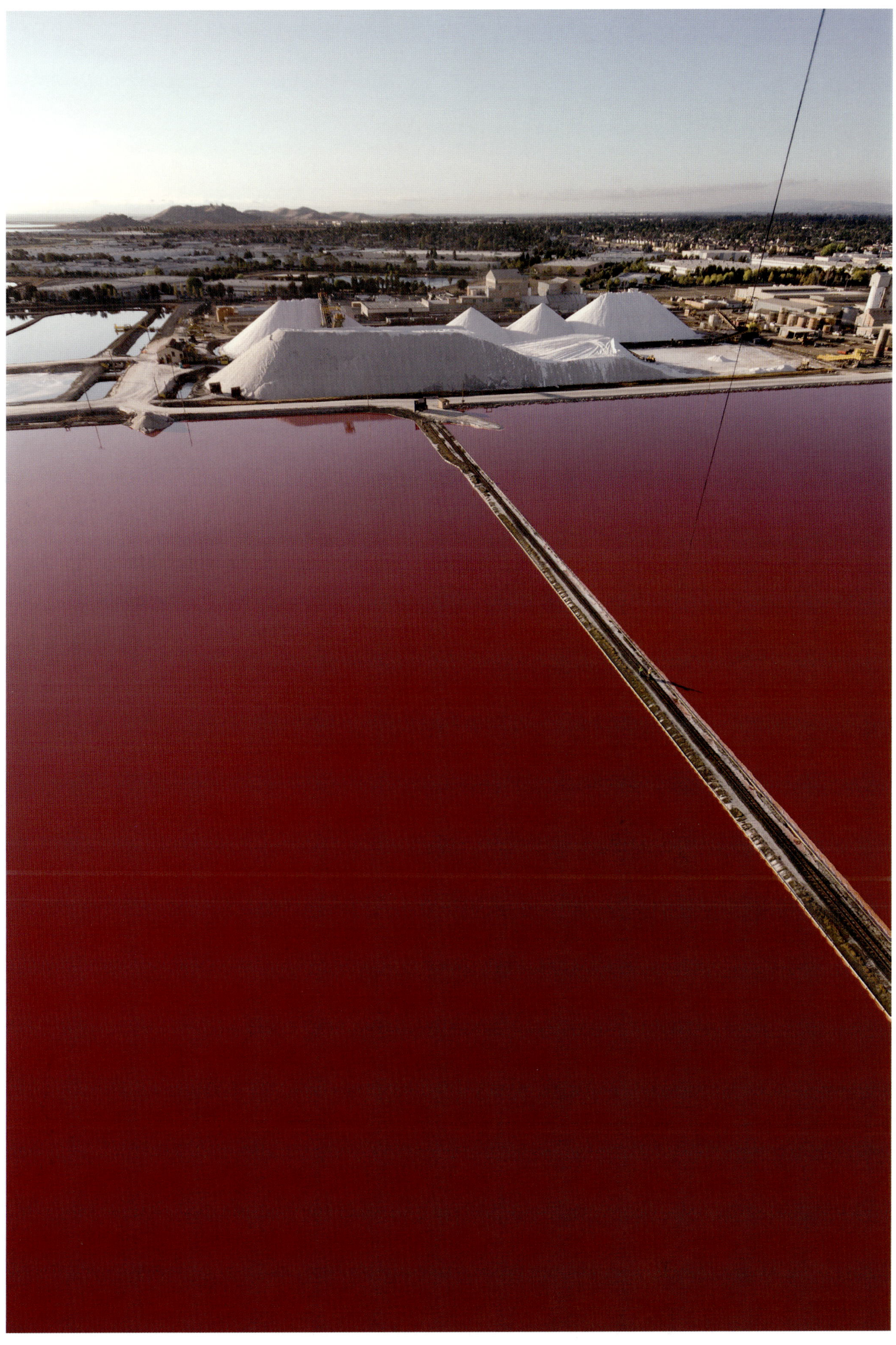

Plate 94. Cargill's Newark facility (Cargill Newark Salt Plant, October 2010)

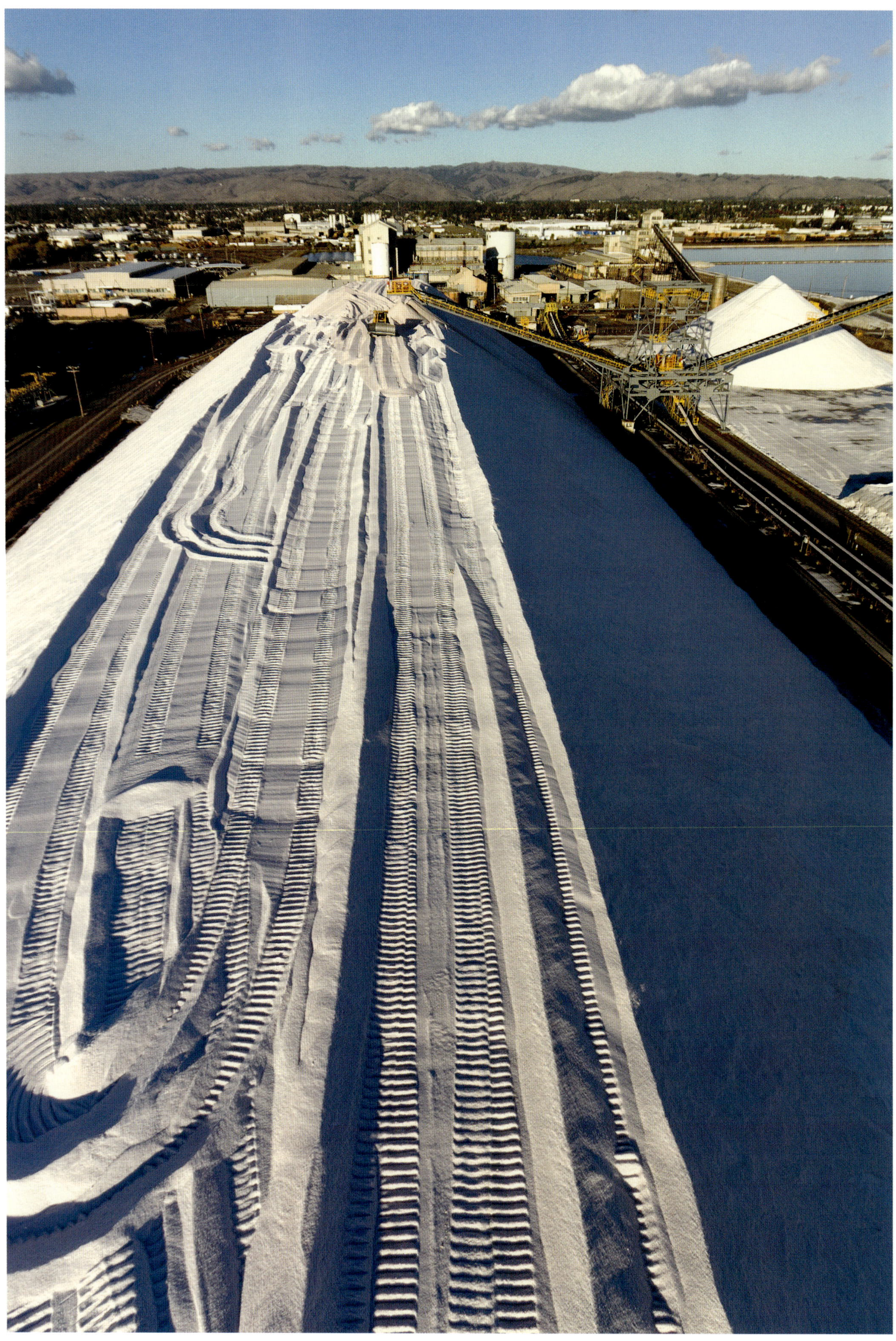

Plate 95. Bulldozer on the stack (Cargill Newark Salt Plant, November 2010)

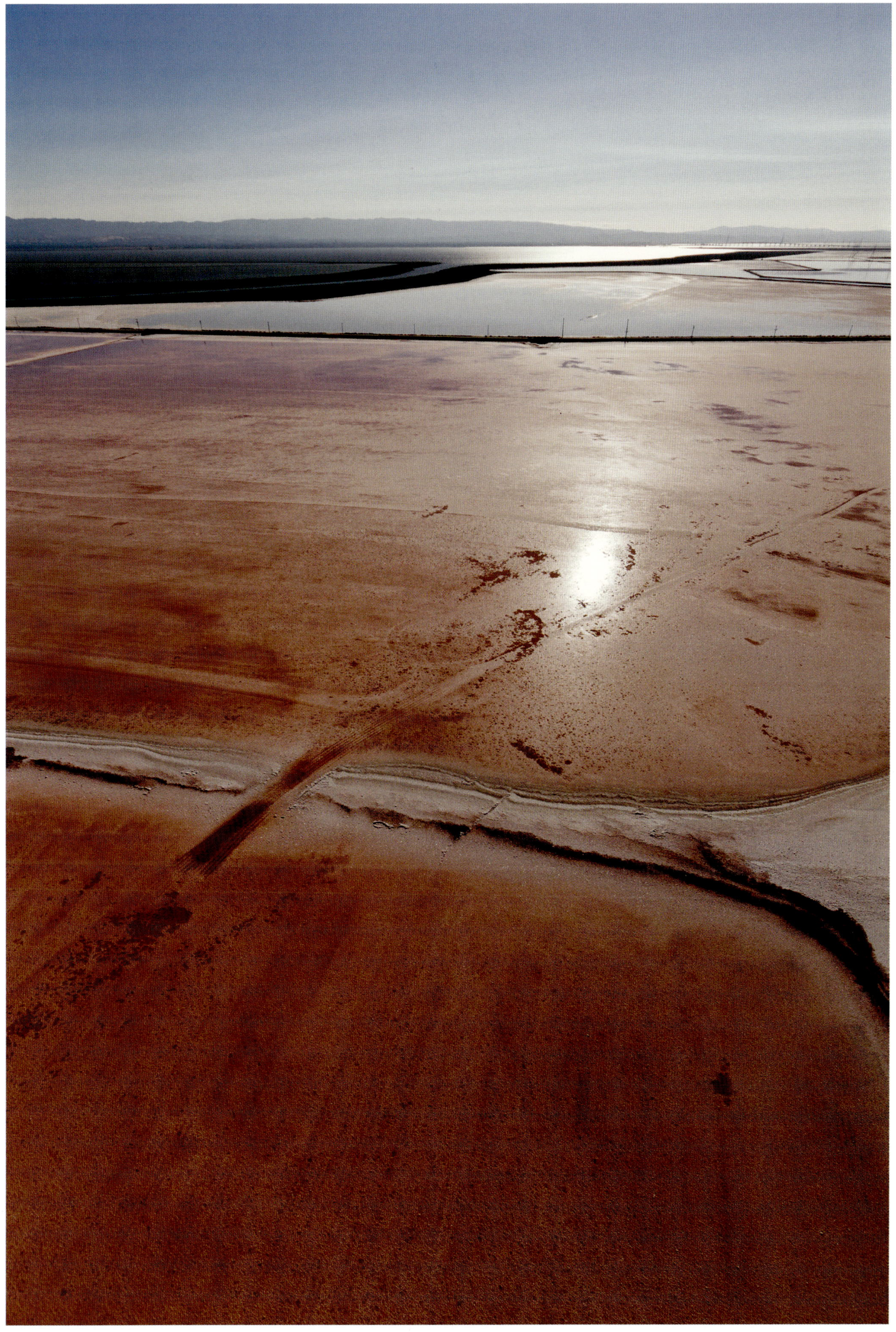

Plate 96. A large expanse of crystallizer bed (Newark Crystallizer Beds, August 2010)

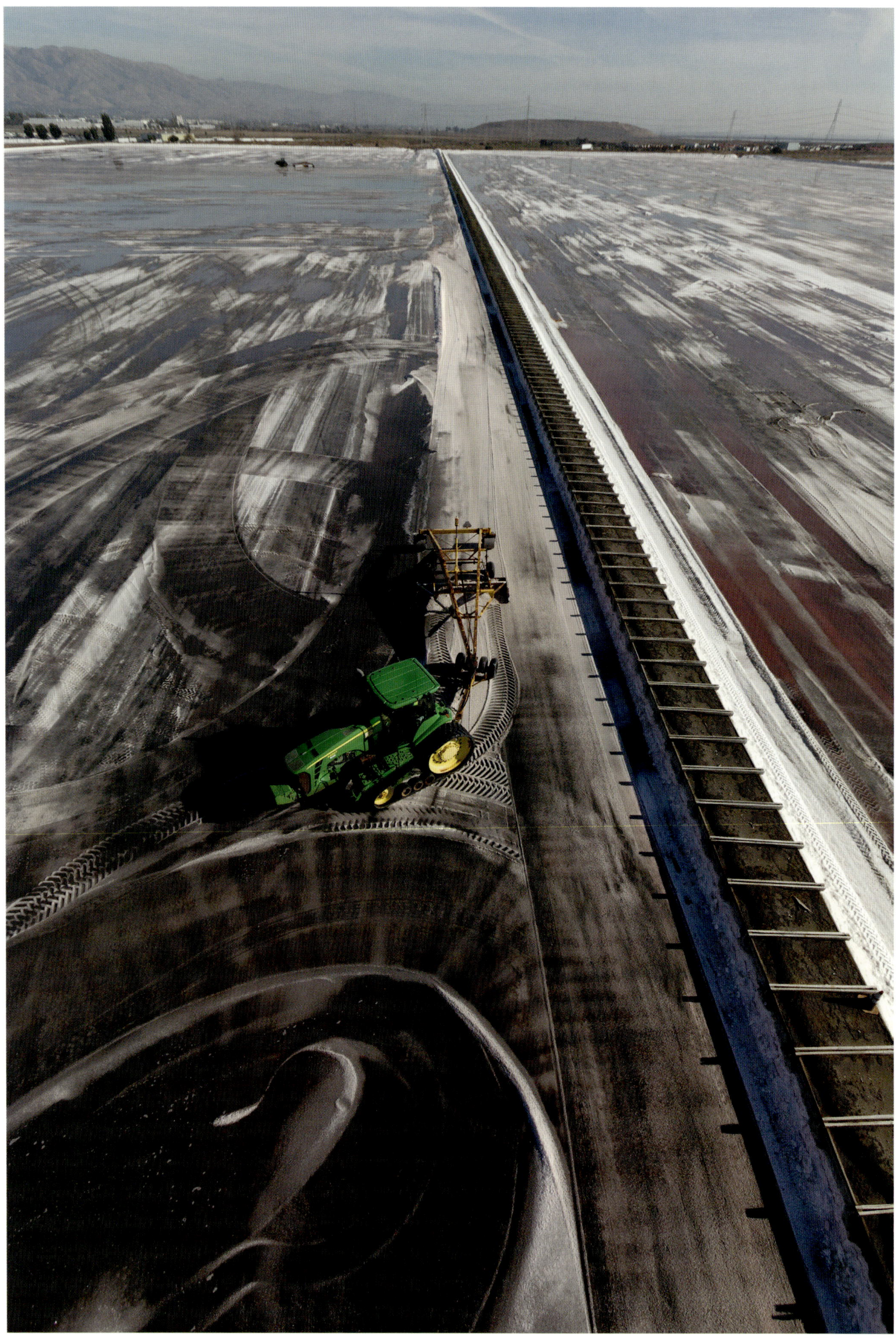

Plate 97. Leveling the floor (Newark Crystallizer Beds, November 2010)

Plate 98. Patterns of harvest (Newark Crystallizer Beds, October 2010)

n many ways the landmark deal in 2003 that started the South Bay Salt Pond Restoration Project expanded Cargill Salt's role in the South Bay. It is notable that Cargill's participation in this agreement was not limited to just the sale of property and mineral rights. They are now active allies in the planning and implementation phases of the restoration project, bringing to the table invaluable expertise in management of the ponds and assets like the *Mallard II* dredge. As some ponds await preparation for a return to tidal flow, Cargill has helped manage them through the interim periods. Meanwhile, they have collapsed their network of active evaporator ponds to a fraction of its former size while maintaining the salt production of previous years—around five hundred thousand tons of salt per year. Almost all of this salt harvest occurs in a single location, the crystallizer beds and salt plant at the former site of Arden Salt Works #2 in Fremont, where Cargill Salt runs an impressively efficient operation.

If John Johnson, the first to dike salt ponds 150 years ago near Eden Landing, could see the current-day Cargill operation he would surely be amazed. The colors and textures of the crystallizer beds would seem familiar but certainly not the scale. Where the harvest was once gathered laboriously with pick and shovel, now large machines grind and gather salt on the crystallizer bed floors. Johnson once struggled with mud contaminating his salt; Cargill has eliminated mud and provided structural support by developing crystallizer beds with a permanent floor of salt. Where Johnson's crew used hand-pushed wheelbarrows, the Cargill harvesters fill massive, balloon-tired dump trucks with thirty-five tons of salt in ninety seconds. During the late fall harvest season, a continuous stream of these trucks hauls salt to Cargill's wash house, where it is cleaned and transported by conveyor belt to interim storage. Johnson's waist-high stacks of harvested salt have now become huge twin stockpiles seventy-five feet high and five hundred feet long, groomed by bulldozers and visible for miles. Once harvested, the crystallizer beds, one of which would contain Johnson's entire works several times over, are now precisely graded for the next year's harvest by laser-guided scrapers. Where Johnson shipped low-quality salt from his boat landing to San Francisco for processing, Cargill's modern, automated factory washes, grades, packages, and distributes salt tailored for different uses, including culinary salt of highest purity.

After a few years I secured permission to take photographs on Cargill Salt's property, an effort well worth the wait. After countless trips to photograph traces of previous salt operations, here at last was access to an active saltworks—dynamic, bustling, purposeful. I was able to visit during harvest season when the crystallizer bed landscape is animated by industrial processes and the flow of huge quantities of salt. The crystallizer beds are the territory of archaebacteria, the extreme halophiles responsible for the most vivid of the salt pond colors. Yet even these colors are rivaled by the textures of salt and machinery, of process and past. This was a fun place to photograph and access to do so, for which I am most grateful, made my South Bay explorations seem somehow whole. The South Bay will certainly have salt ponds for the foreseeable future. They serve as incubators for important segments in the early stages of the food chain while providing specialized habitat that would be difficult to replace. As the South Bay Salt Pond Restoration Project assumes stewardship of large tracts, Cargill Salt's reduced holdings will continue salt industry traditions while providing wildlife habitat that complements the restoration efforts.

In the late 1990s the possibility of Cargill Salt's sale of property and mineral rights offered a tantalizing opportunity to consider large portions of the South Bay a tabula rasa, at least in terms of future agenda. Preparatory work examining the feasibility of restoration, notably the 1999 *Baylands Habitat Ecosystem Goals Report*, served as a rallying point for restoration advocates. It had taken vision, ingenuity, technical expertise, and decades of sustained effort to transform the South Bay marshes into the salt-pond landscape of the day. Cargill Salt, like their predecessors, had developed clever strategies to harness the considerable powers of

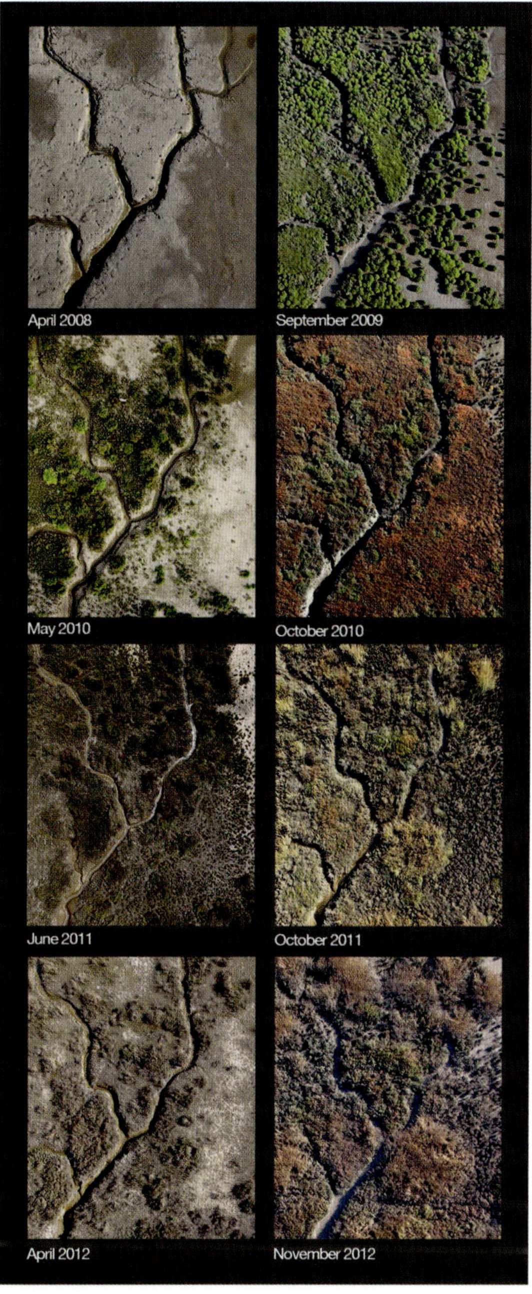

Sediment, and then new vegetation, transform a former salt pond after a 2006 levee breach (Salt Pond A21).

nature in producing salt. It became clear that restoration efforts would require a similar commitment of dedication, knowledge, and time as well as a reliance on natural processes using techniques yet to be fully developed. But the payout envisioned was huge—large sections of former South Bay wetlands restored to tidal flow, a diversification of wildlife habitat with attention to species of concern, and the reclamation, at least in part, of environmental qualities thought lost forever. The potentials of a South Bay restoration were within grasp if not the actual methods.

The alliance of environmental advocates formed in the prospectus stage of the restoration project has continued through acquisition, planning, and implementation. The California Department of Fish and Wildlife has joined the US Fish and Wildlife Service, managers of the Don Edwards

San Francisco Bay National Wildlife Refuge, as a direct manager of restoration properties. The California Coastal Conservancy provides overarching project management while the US Geological Survey leads science projects supporting the restoration effort. These core members of the project management team are joined by literally dozens of other government agencies, private foundations, and citizen groups in implementing what is considered a fifty-year, multibillion-dollar restoration effort. Cargill Salt serves as a knowledgeable and cooperative neighbor to the effort, lending expertise and tools for managing the wetlands.

The project has three large units of activity: the Eden Landing Ecological Reserve, just south of the eastern end of the San Mateo Bridge; Ravenswood, at the western end of the Dumbarton Bridge; and Alviso, at the southern end of the bay. Management goals for these sites have coalesced around the major objectives of establishing a more natural mix of tidal wetlands and managed ponds for wildlife habitat, providing flood control for adjacent communities of humans, and offering opportunities for outdoor recreation. Project planning was informed by prior experience in small-scale Bay Area marsh restoration projects, the 1990s large-scale salt-pond restoration project in Napa County, and consultants aplenty. From the beginning it was understood that the restoration process would have to address difficult issues including pond subsidence, sources of sediment, desalinization, contaminants (especially methyl mercury leached from former cinnabar mines in the south Santa Clara Hills), invasive species, species displaced from decommissioned salt ponds, and the effects of sea level rise. These challenges are being met with an approach that combines rigorous scientific investigation, ongoing consensus building, and adaptive management.

Adaptive management is more than a slogan in the South Bay Salt Pond Restoration Project. Particularly well suited to the project's long time frame and multiple uncertainties, adaptive management is a structured, iterative process that refines decision making with feedback loops from experimentation and information gathering. In the South Bay it involves testing alternative environmental interventions, learning from their performance, and adjusting strategy on the fly where indicated. The process is supported by an impressive range of science projects examining hydrology, sedimentation, contaminant dispersal, species behavior, and more. The project will ultimately convert large zones of former salt ponds into interconnected habitats—tidal marsh, mudflats, seasonal wetlands, managed saline ponds, and adjacent buffer zones—with a majority of these having tidal flow.

My early South Bay photographs were taken under permit from the Don Edwards San Francisco Bay National Wildlife Refuge. In 2008 my permits expanded to include the entire South Bay Salt Pond Restoration Project, providing a box seat for the opening acts of the restoration process and an opportunity to contribute to it. While the project is still in its early stages, it has been exciting to see nature reacting to the relatively bold changes underway in the South Bay. In recent years I have developed a mild sense of urgency in photographing the area, for the arrival of each new feature in the restoration—a flooded pond or newly created marsh—marks the departure of the landscape feature it replaces. From visit to visit I see tangible change in the landscape and am struck that landscape features photographed just months ago have now disappeared for good, all part of starting the next chapter.

In the Alviso Unit I visit several areas regularly. In 2006 the salt ponds flanking the abandoned hamlet of Drawbridge (A20, A21) were the first returned to tidal flow by the restoration project. At that time the ponds' plateaus were barren plains of deposited gypsum. Since 2008 I have been visiting the Drawbridge ponds twice a year to photograph quite noticeable changes as sediment is deposited, the organic patina of mudflat appears, and vegetation takes root. To the south I photograph regularly along the railroad corridor established in 1880, an area where ponds are being repurposed for avian habitat. Flanking Alviso there are the entertaining textures of the New Chicago Marsh to the east, an area with tenuous connections to tidal flow, and to the west a large salt pond recently connected to the tides via an engineered flow control structure. All eyes are on the progress of this pond (A8), as it is a test case for the management of methyl mercury in sediment deposits. Further west is the fascinating Knapp Tract, a former salt pond that was home to over twenty thousand pairs of nesting California gulls until it was returned to tidal flow in late 2010. With a return to the tides its bottom, formerly quite graphic as fallow salt ponds go, quickly attained the smooth surface of a mudflat, and the gulls have moved on to settle elsewhere.

When I first visited the Eden Landing Ecological Reserve in 2008 it was a landscape composed of large disused salt ponds stretching from Highway 92 to the Old Alameda Creek channel. In this popular area for duck hunters the northern half of these ponds filled with rain and bay water in the winter, then dried out in the summer. Mount Eden Slough itself, the organizing channel of early saltworks, was dammed and had ceased to exist past the former site of Eden Landing. To the east, adjacent to the Eden Shores housing development, was the North Creek Marsh, a former saltworks site returned to tidal flow in 2005. I visit Eden Landing often to photograph, in part due to interesting traces of the past like the Oliver Salt Works ruins and in part due to the actively evolving landscape as restoration efforts progress. In recent years managers have worked carefully to desalinate the area's former salt ponds without harmful discharge into the surrounding environment. During this period of desalinization, disused salt ponds with high residual salt offered entertainingly colorful scenes. The same areas have since adopted the muted colors of lower salinities and are now acquiring the surface textures associated with tidal flow. The dam on Mount Eden Slough was removed in 2010, connecting large areas on the east side of the reserve to the tides and the prospect of becoming marsh again. More recently, levees along the Alameda Creek Flood Control Channel were breached to connect large areas near the bay shore to tidal flow for the first time in over a hundred years. With each passing year I find aspects of the Eden Landing Ecological Reserve changed—familiar features gone, nascent wetlands emerging—signs that make restoration progress excitingly tangible.

The Ravenswood Unit has received short shrift in my work, the consequence of nearby airports and associated low-flying planes. In Ravenswood I have photographed the most ambitious of the restoration project's habitat experiments, a tiny archipelago of nesting islands created south of the Dumbarton Bridge approach. Here a former salt pond (SF2) was rebuilt with substantial flow control structures to establish mudflat and nesting habitat for shorebirds. Another portion of the site offers dry pond bottom in the service of snowy plovers. The birds are still thinking about it. I have photographed often to the north at Bayfront Park in Menlo Park, which adjoins both marsh and an interesting set of fallow salt ponds.

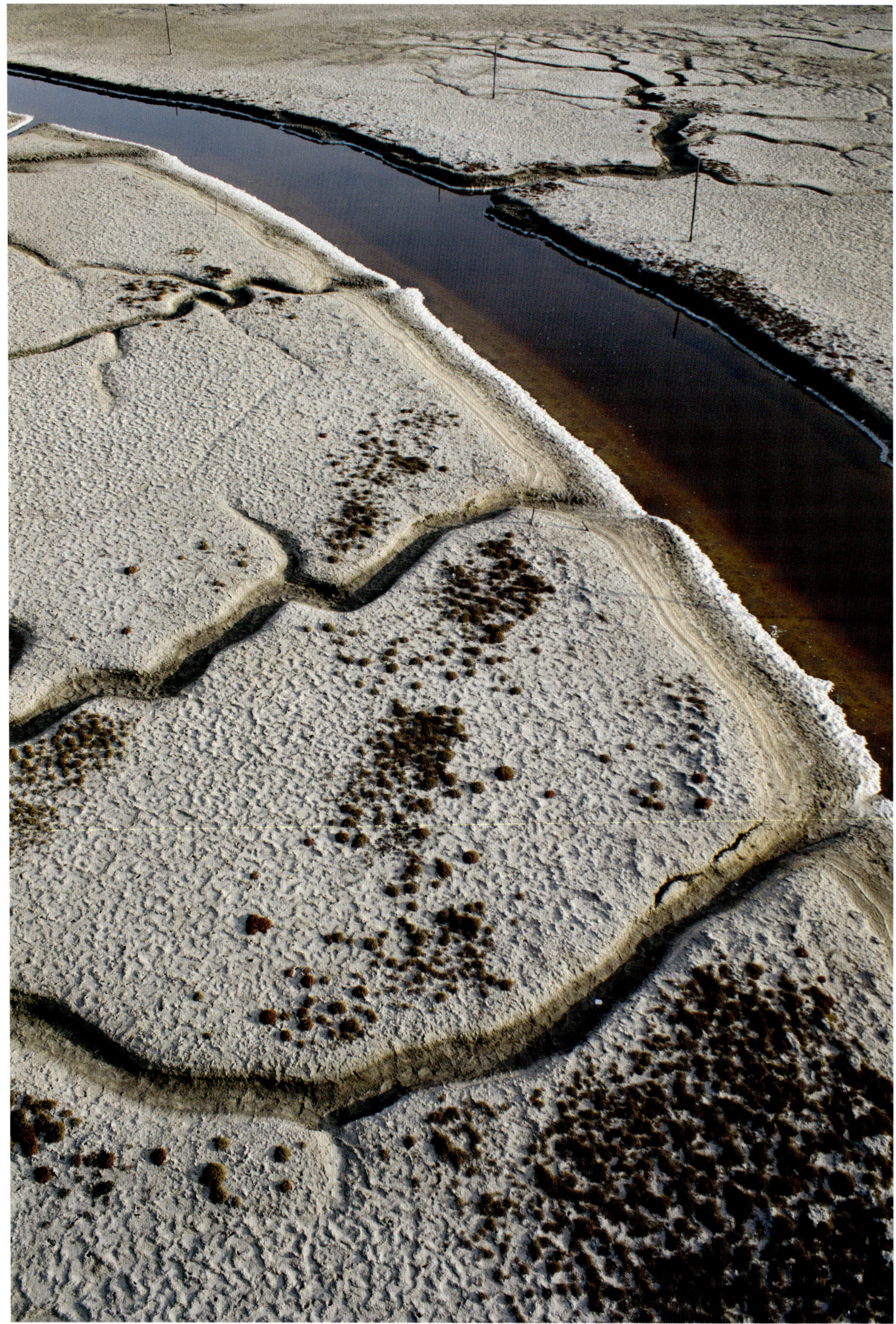

Plate 99. Pond surface before renovation (Salt Pond SF2, December 2008)

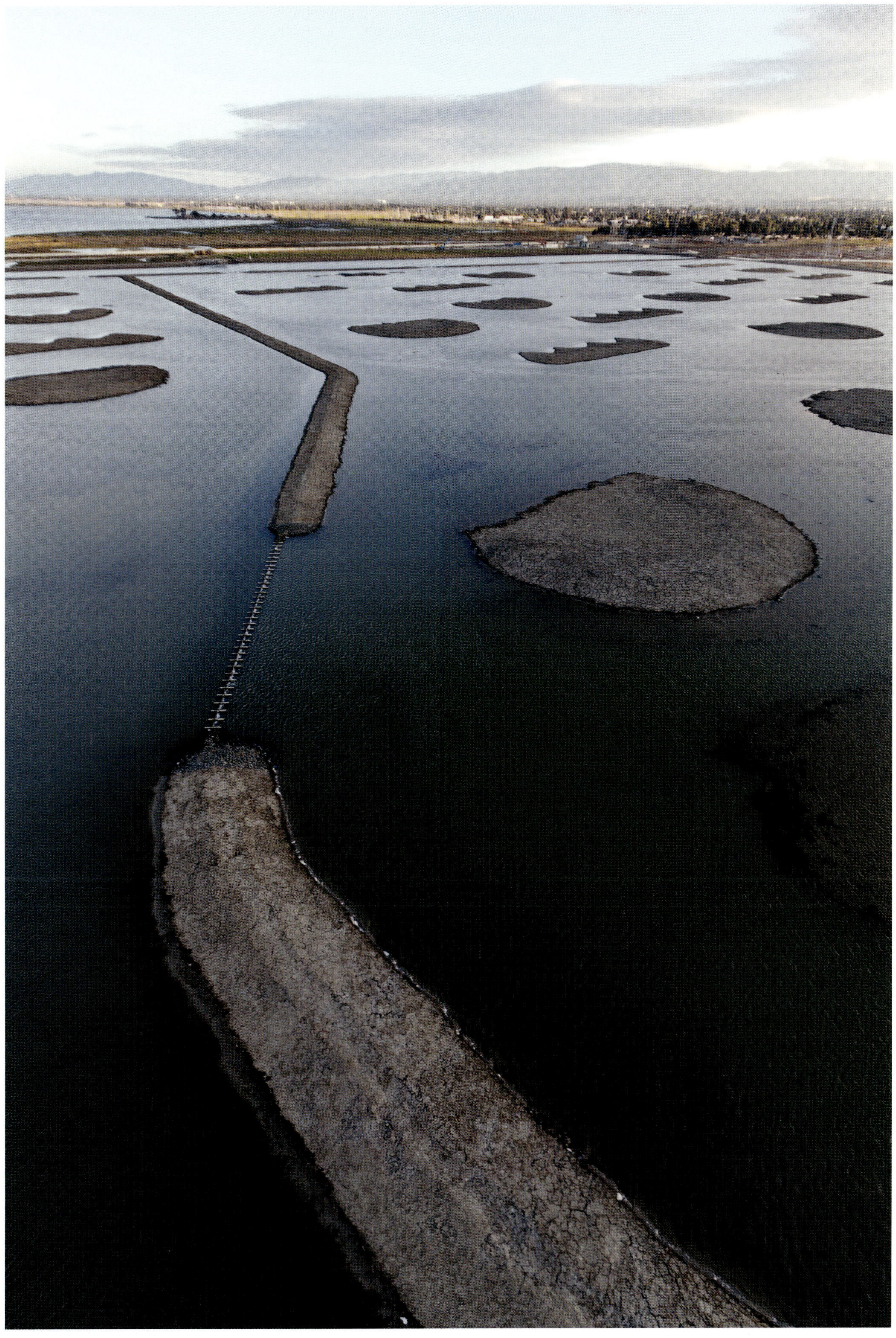

Plate 100. An experiment in avian habitat (Salt Pond SF2, August 2010)

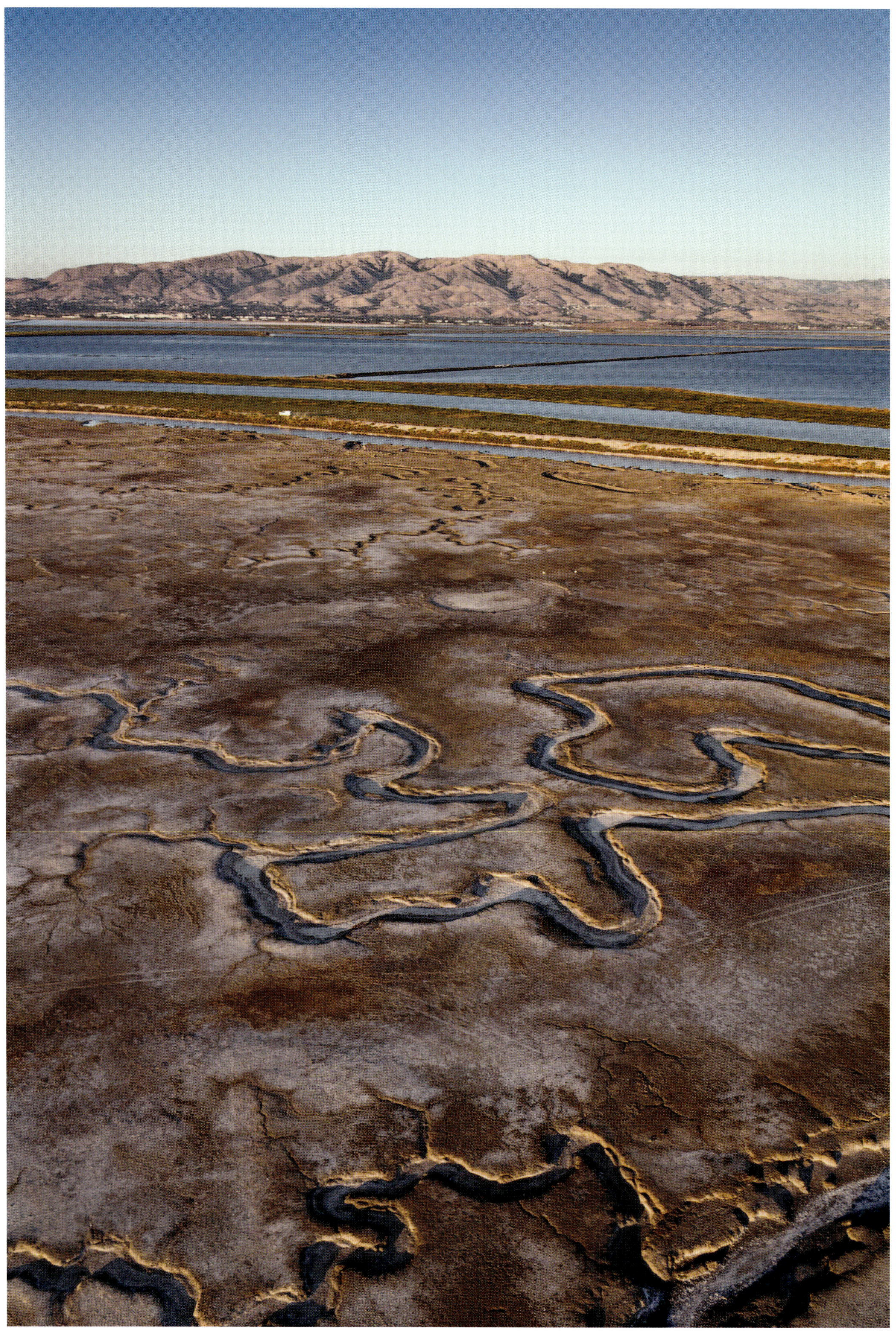

Plate 101. Pond bottom before renovation (Salt Pond A6, August 2010)

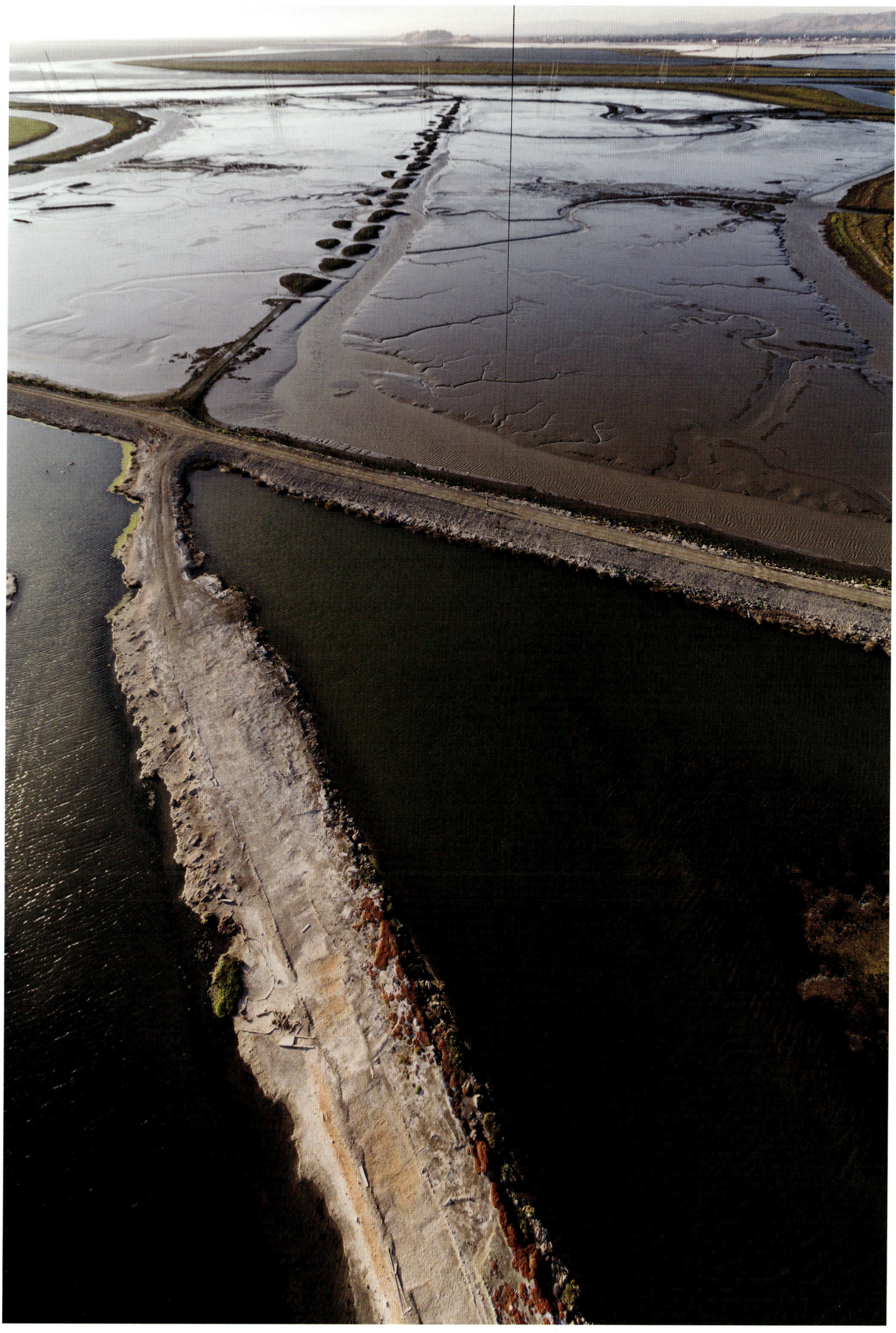

Plate 102. Mudflats forming after the breach (Salt Pond A6, August 2011)

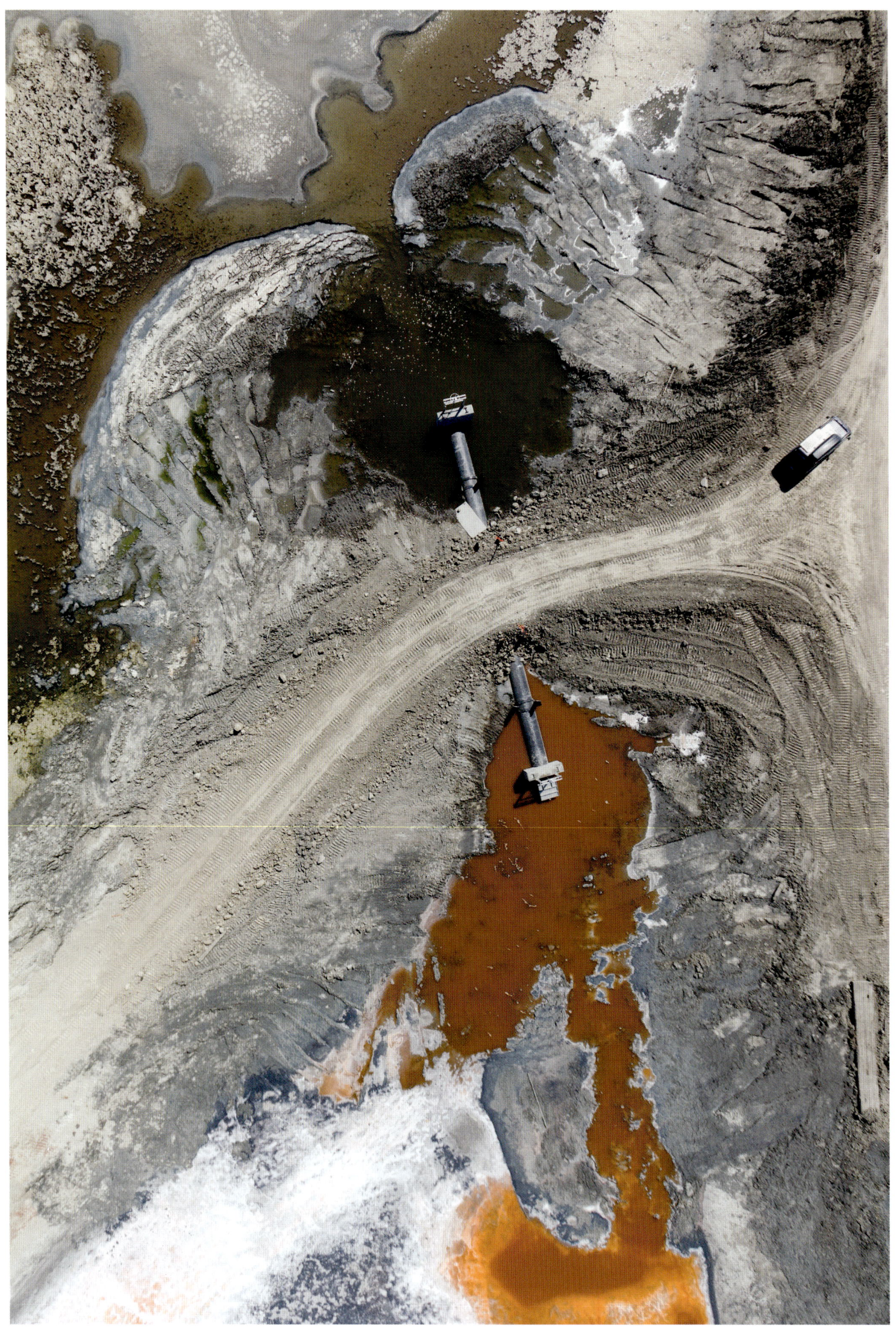

Plate 103. Newly installed flow control gates (Salt Ponds E13 and E14, August 2011)

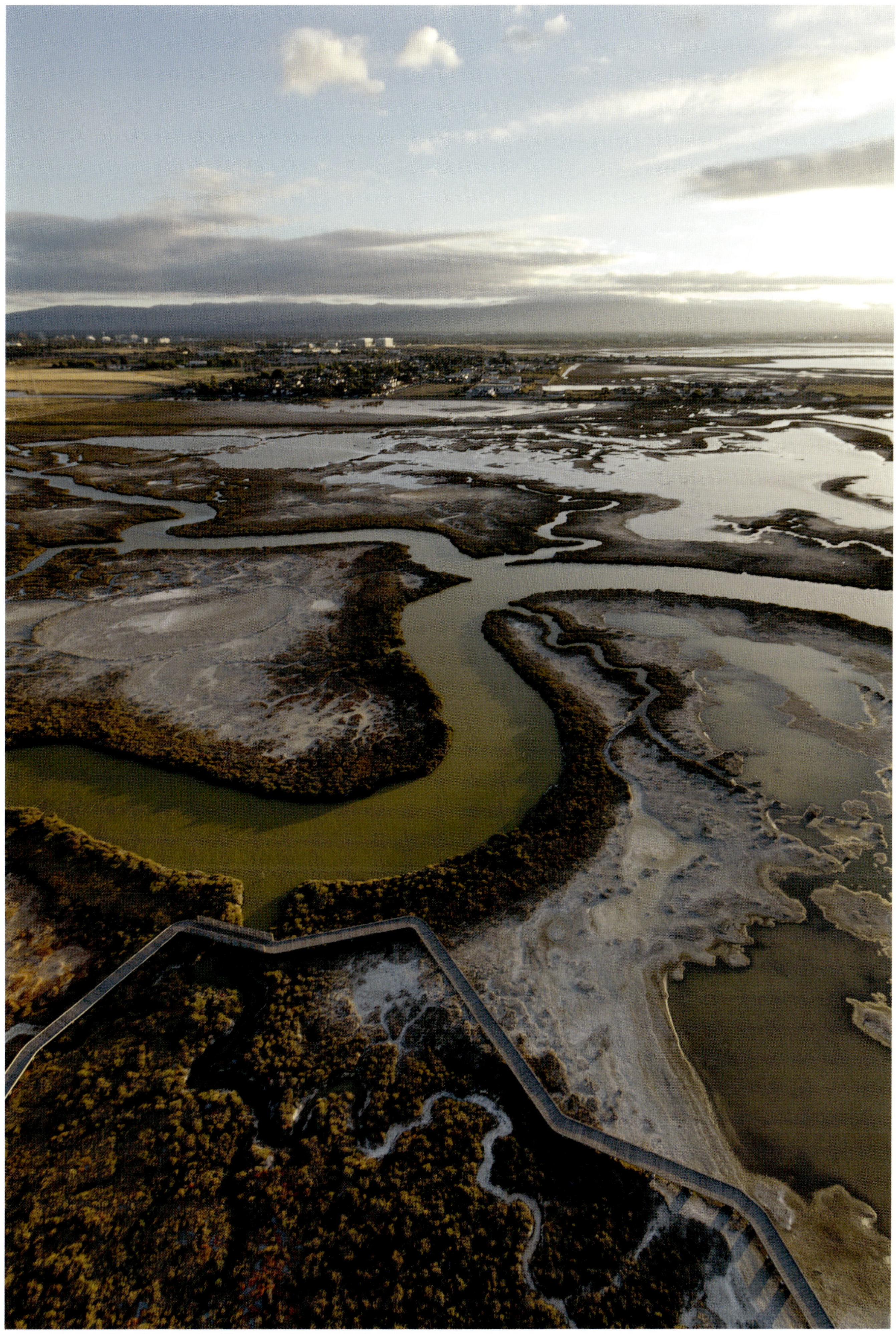

Plate 104. Alviso in the distance (New Chicago Marsh, September 2010)

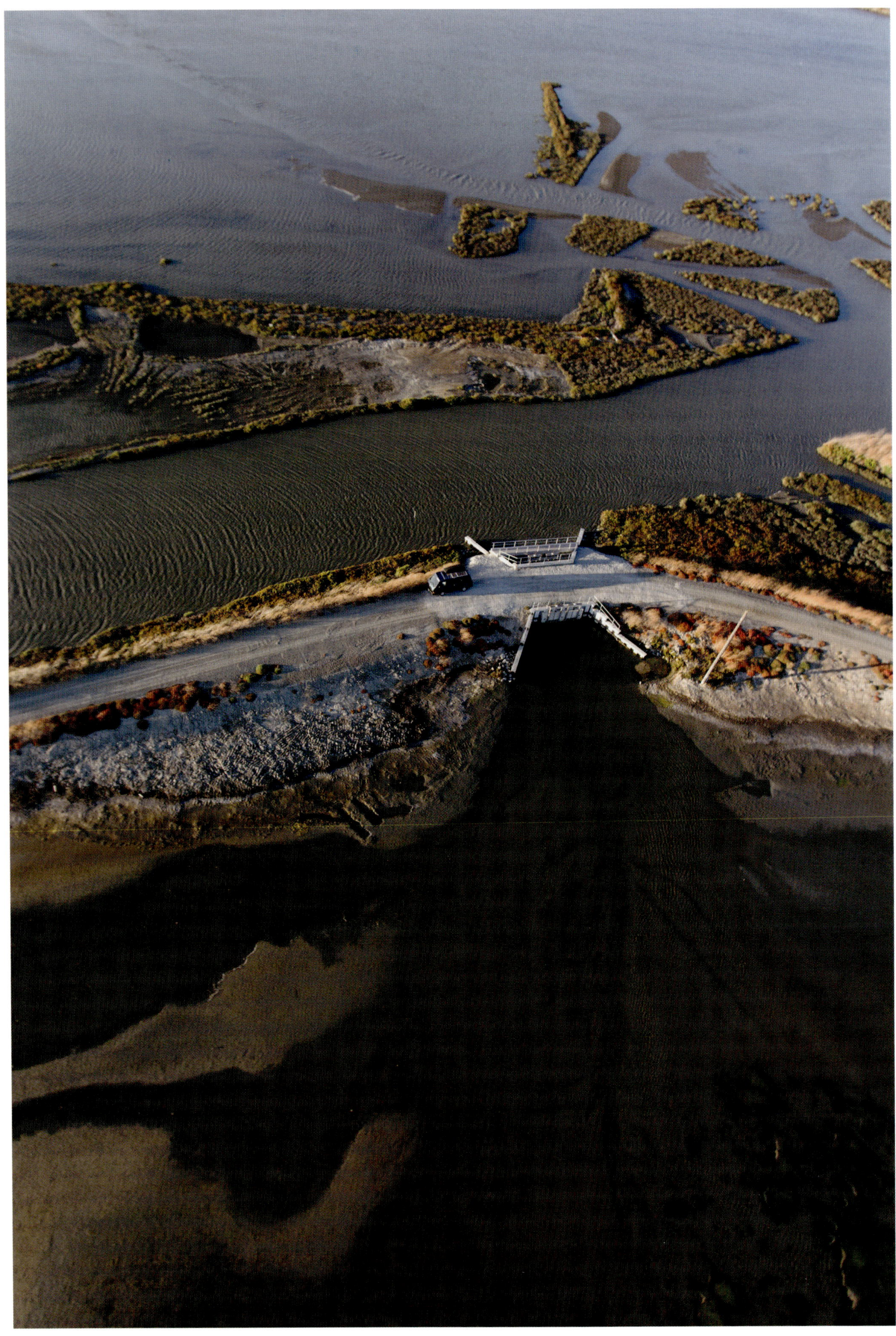

Plate 105. Newly installed flow control gate (Salt Pond E12, September 2012)

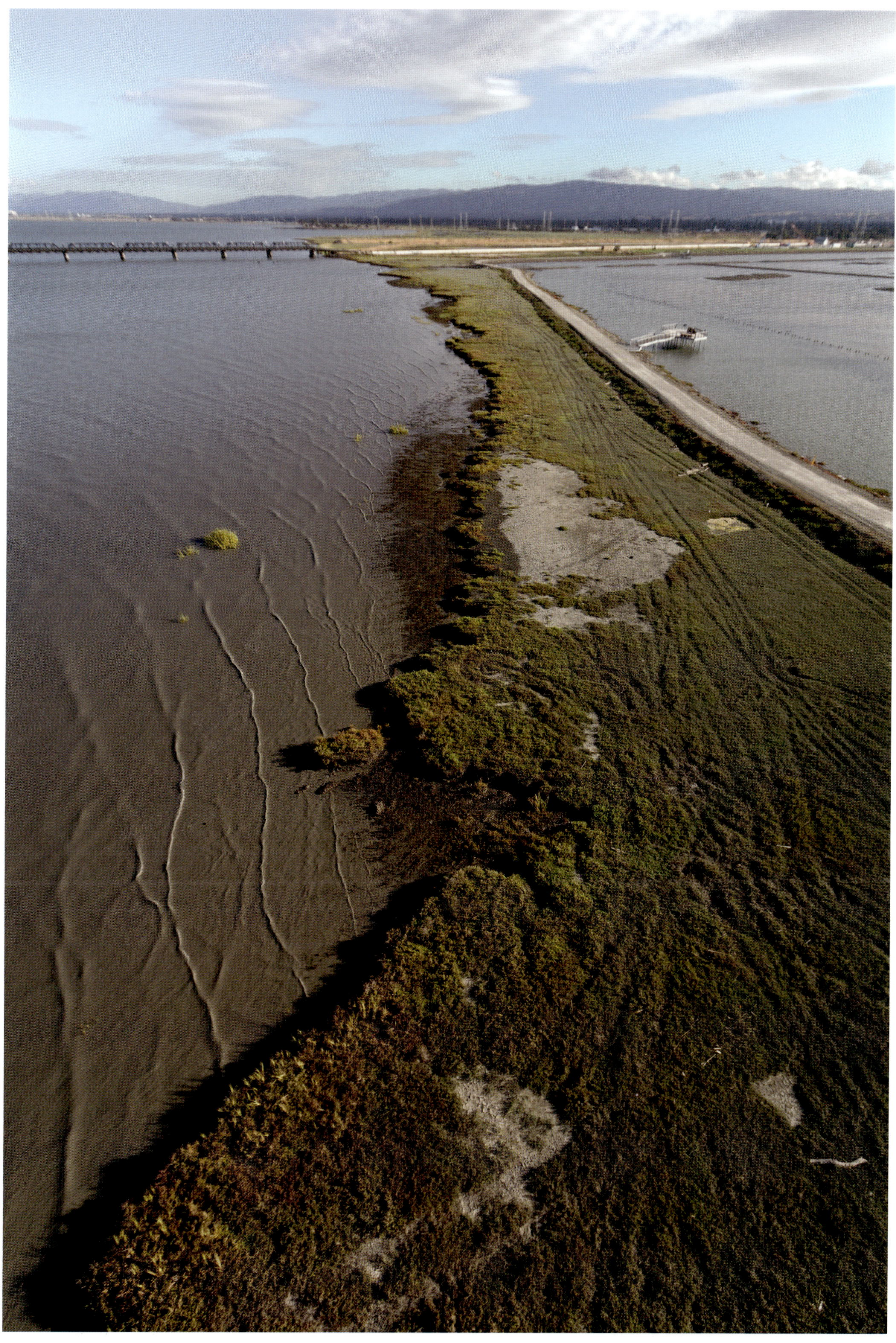

Plate 106. Trail along edge of bay (Bay edge of Salt Pond SF2, August 2011)

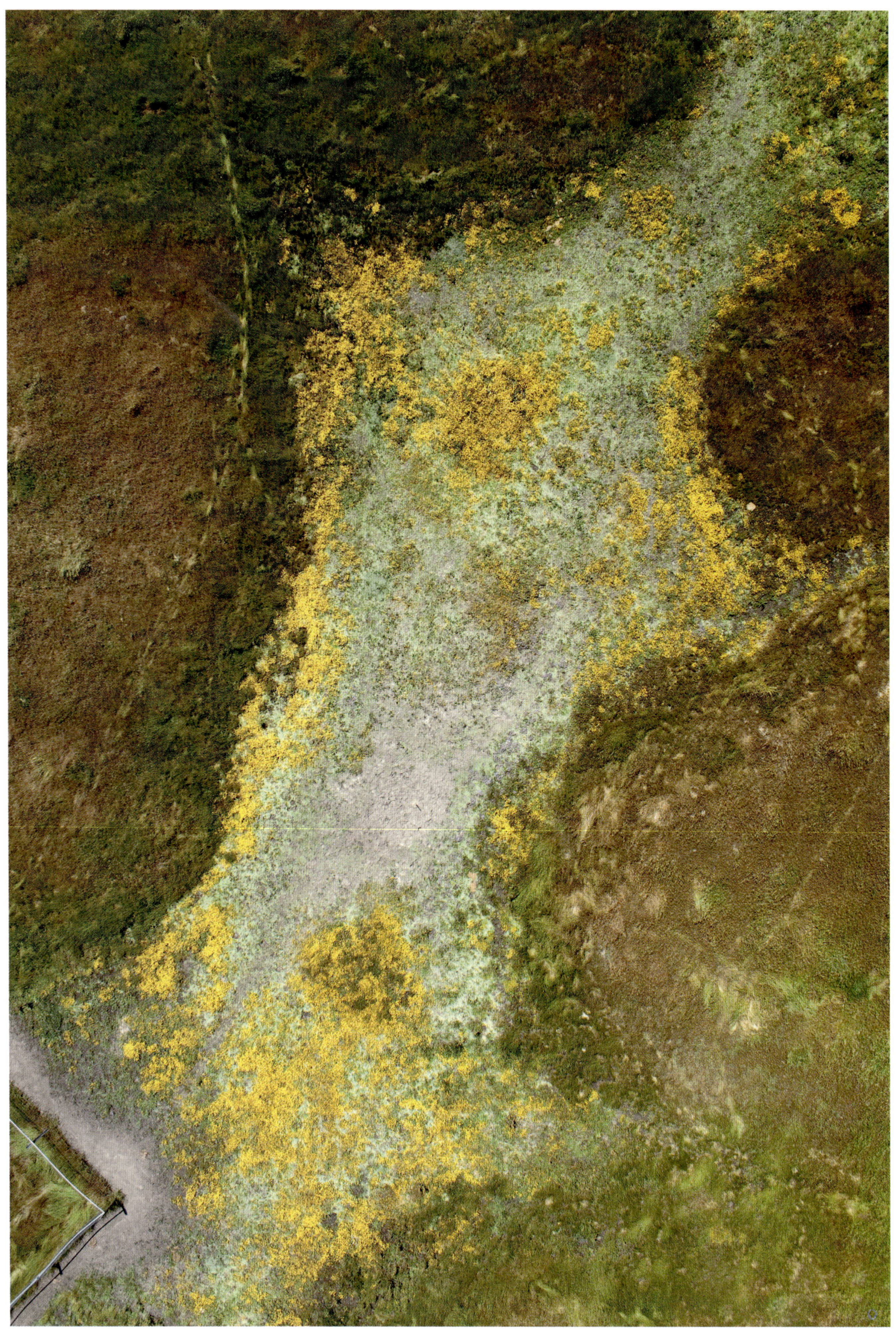

Plate 107. Contra Costa goldfields, downingia, and woolly marbles (Warm Springs Seasonal Wetland, May 2010)

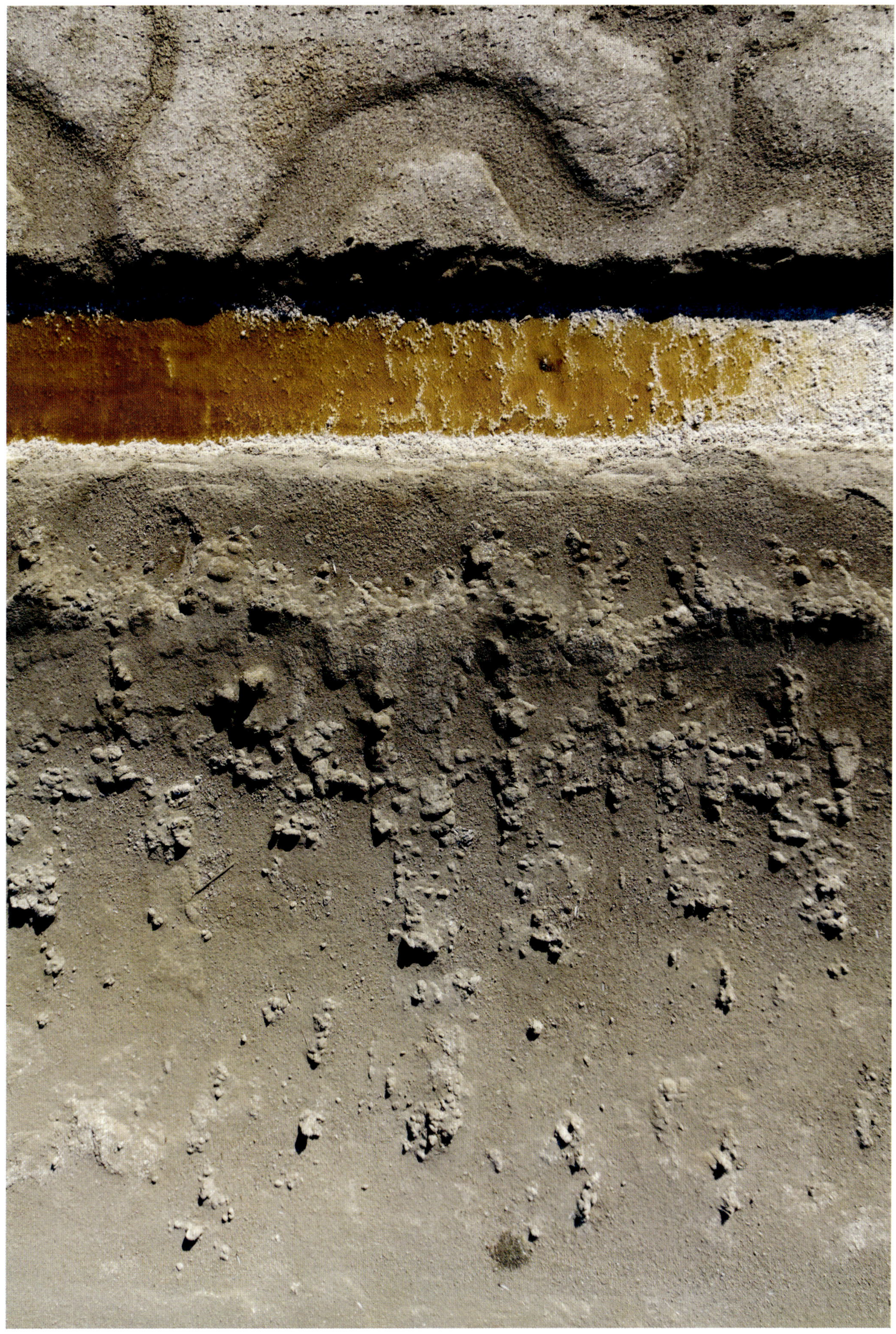

Plate 108. Preparations for a 2011 breach (Salt Pond A6, August 2011)

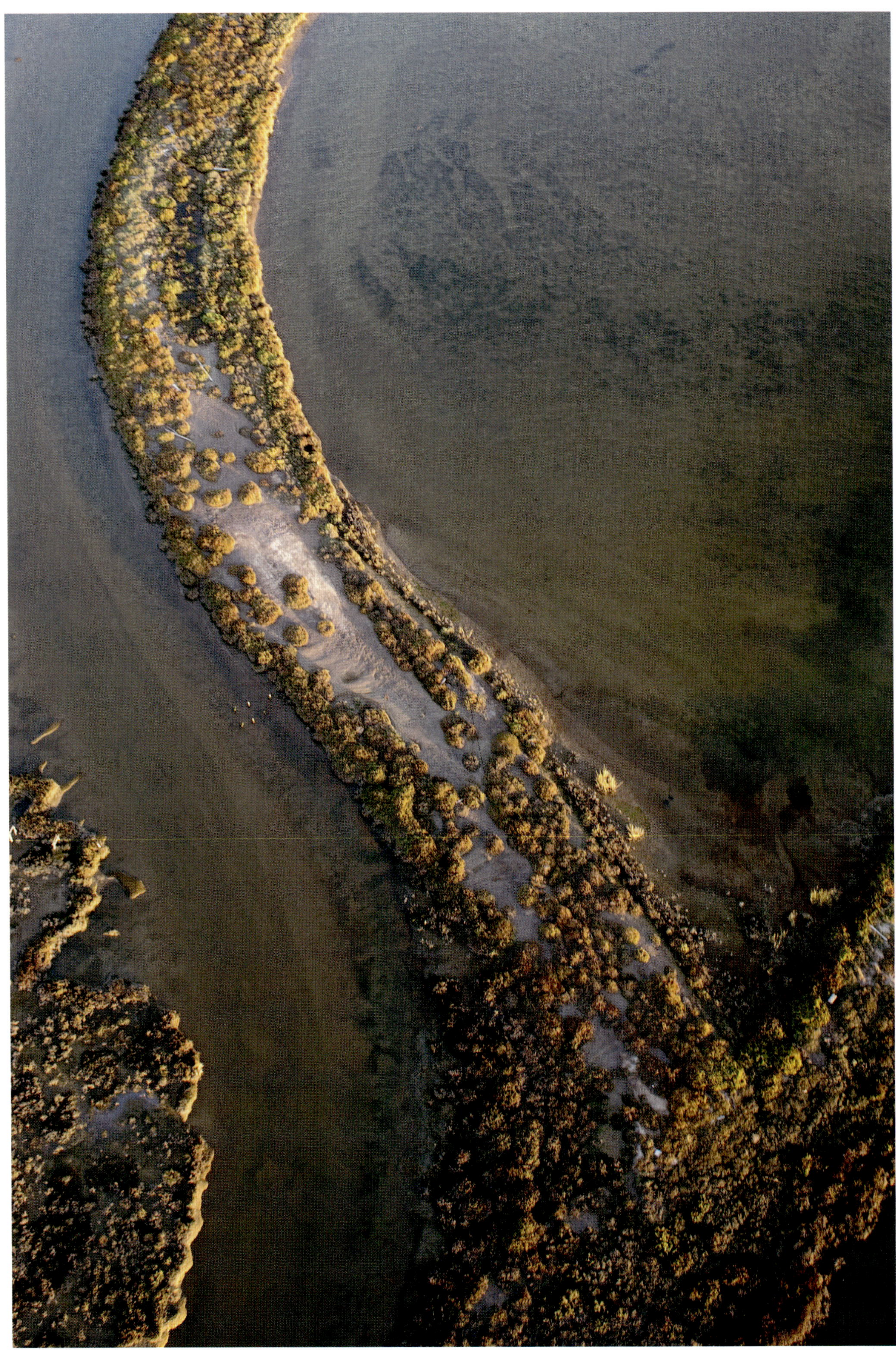

Plate 109. Curved levee (North Marsh, September 2009)

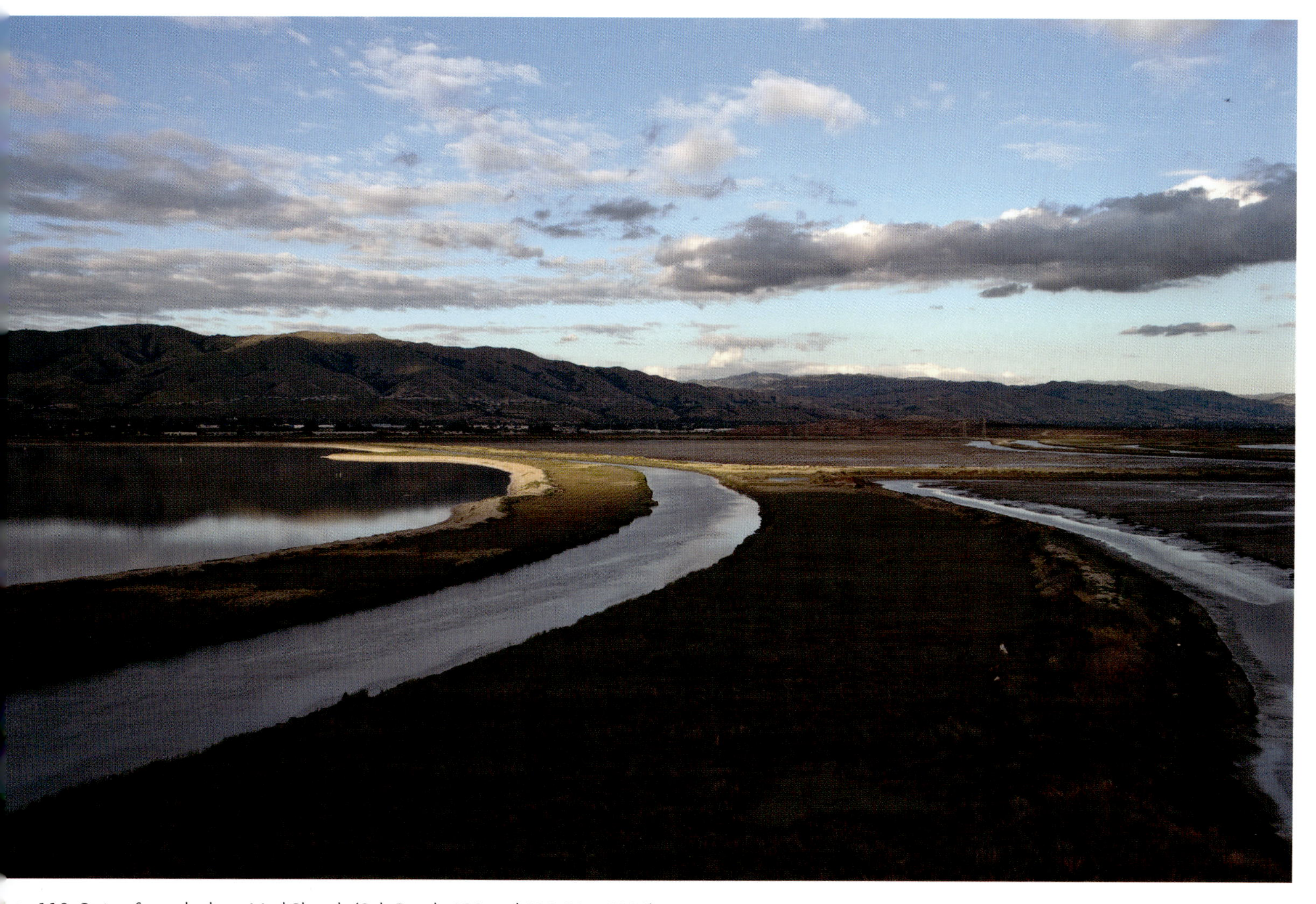

ate 110. Strip of marsh along Mud Slough (Salt Ponds A23 and A20, May 2010)

s my experience in the South Bay grows, so does my appreciation for the resilience of nature.
I would never have imagined that the stark, monotonic extremes of a salt crystallizer bed
could somehow be softened enough to support the intricacies of marsh life. Yet others
did imagine this and furthermore demonstrated, in places like Fremont's LaRiviere Marsh, just
such a restoration. This imagination, this optimism, has led to a sustained, rigorous program
for wetland restoration that reaches every corner of the South Bay. For almost a century, the
former South Bay wetlands were an exceedingly salty place. Now thousands of acres in the
restoration units are becoming something else, places with tidal flow and normal bay salinities.
Perhaps large portions will return to their original marsh conditions or nearly so, what wetlands
expert Phil Williams terms "second growth marsh." Other areas will become hybrid landscapes, a
blending of past epochs with whatever course nature chooses following the restoration project's
interventions, or they will assume the mantle of habitat purpose-built for species of concern like
the snowy plover. In every restoration scenario, natural forces play the dominant role and wildlife
populations provide a barometer of success.

In the few years that have elapsed since I became curious about the South Bay I have enjoyed
countless hikes through this landscape and pleasant hours spent deciphering its patterns in my

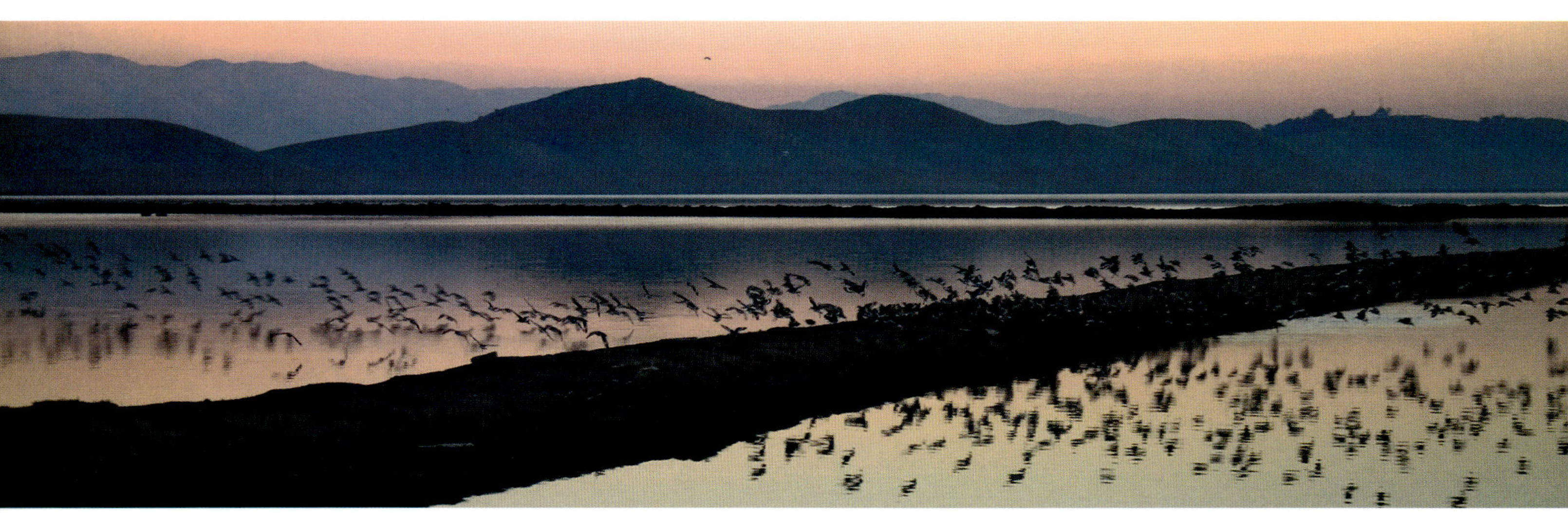

At dusk the South Bay landscape takes on a timeless quality as birds of many feathers settle into
their roosting places across the vast, flat landscape.

photographs. The story in these images is a remarkable tale of transitions—wild frontier to metropolitan center, natural wetlands to industrial ponds, boats to trains to trucks—with each transition leaving its mark on the landscape. And now the story of our day, the South Bay Salt Pond Restoration Project, is appearing in my images. This project, which seemed a conceptual and somewhat abstract endeavor when I began my photographs, is now visibly evident in my aerial views. The Eden Landing Ecological Reserve is experiencing a metamorphosis as salt-pond colors and textures yield to an emerging marsh palette. With each passing year the breached ponds of the Alviso Unit accumulate more sediment and plant growth while the project management team accumulates more knowledge about the natural systems in play, lessons that inform new initiatives. This feels like progress! While there are considerable challenges to overcome, there is also great comfort in observing the motivated, skilled people on the project's front lines. Seeing this work in progress has been a privilege and I will eagerly continue my photographic record as the South Bay's new chapter unfolds.

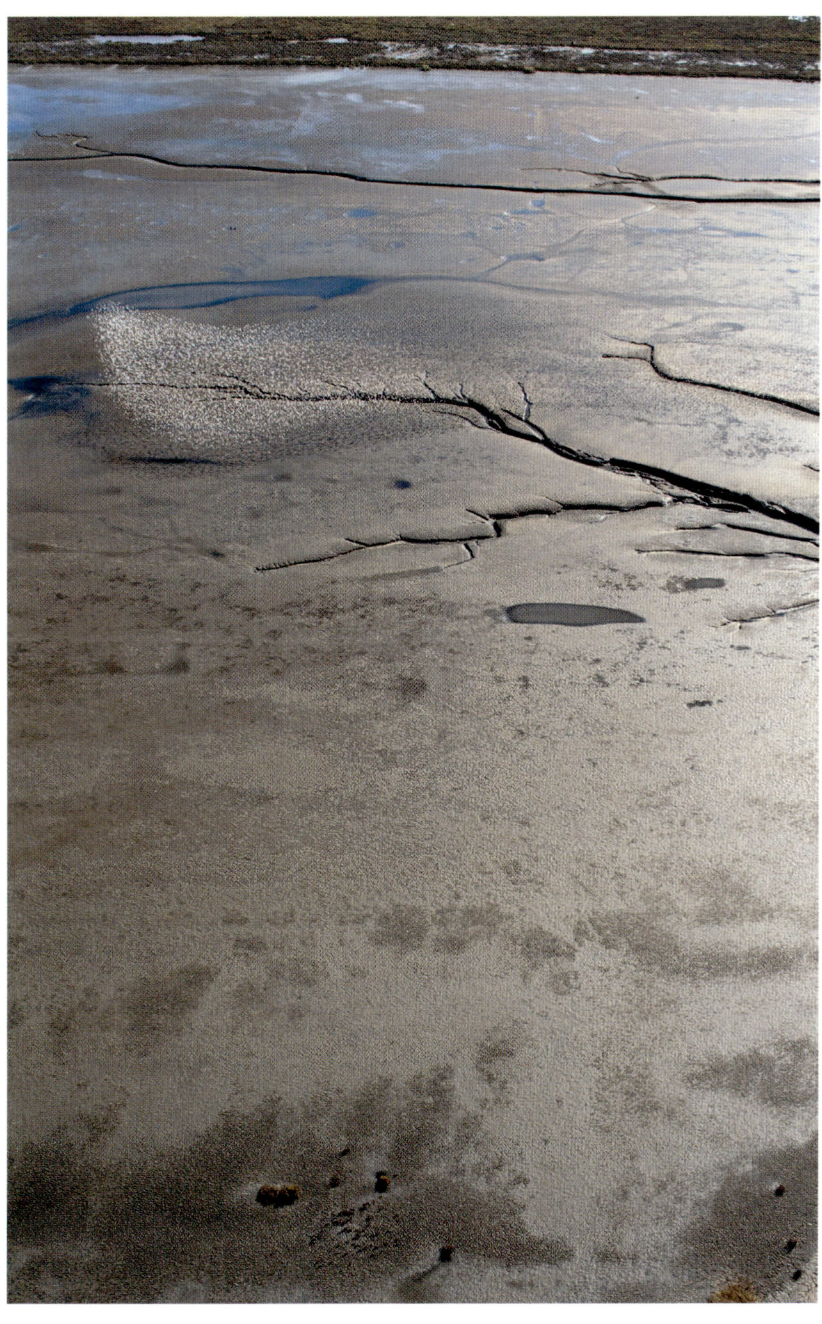

Shorebirds swarm above a reestablished mudflat. (Mt. Eden Creek mudflat, June 2012)

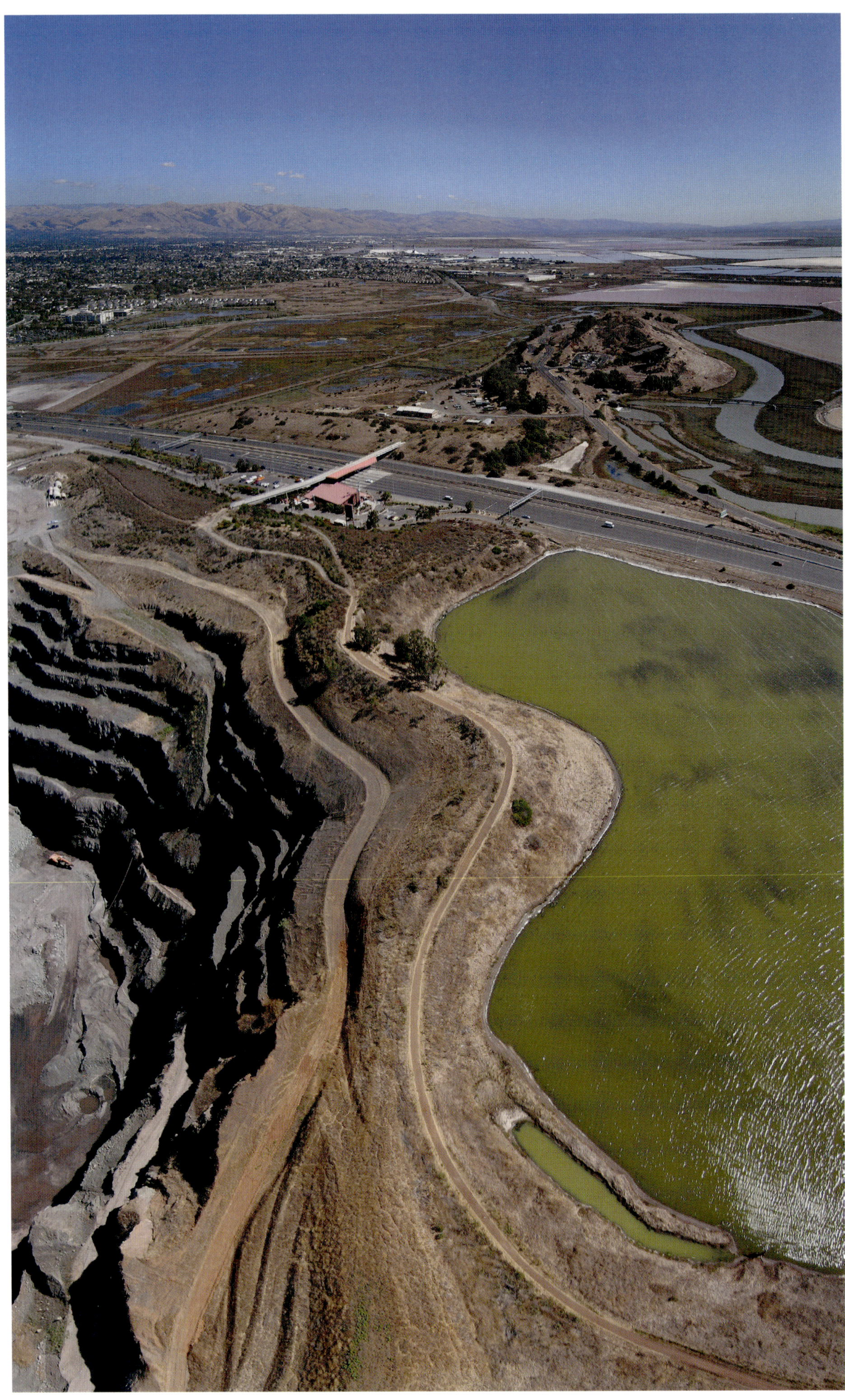

The Red Hill Gravel Quarry, immediately north of the Dumbarton Bridge Toll Plaza, was active until it reached its maximum permitted depth in 2007. The excavation is now being filled in preparation for a new role as parkland.

APPENDIX

South Bay Resources

This is an extraordinary time to explore the South Bay. On one hand, hikers are graced by expanded physical access to a once private landscape; on the other, ample background information is available to inform the armchair explorer. With a modest computer and the connectivity afforded by a coffee shop, one can easily retrieve archival maps, government agency research documents, and high-quality aerial surveys describing South Bay resources that once required access to a university-grade research library. This appendix, while neither exhaustive nor comprehensive, offers a few notes on physical access and pointers to reading material I've found useful.

Hikes

In the 1950s virtually all of the South Bay was private and posted, with access limited to employees of the salt industry, utility workers, and duck club hunters. As the decades passed, public access improved with the creation of new parks, the Bay Trail Project, and now the South Bay Salt Pond Restoration Project. The Association of Bay Area Governments maintains a useful set of trail maps illustrating points of public access in the South Bay.

While public access to the South Bay landscape has improved, please note that large areas are closed to the public for the protection of wildlife habitat. My special use permits granting access to normally off-limit areas for photography come complete with pages of exacting rules, restrictions, and blackout dates designed to protect wildlife and sensitive areas. Please respect signage that designates areas as off limits. Happily, there are many areas the public can visit along the entire perimeter of the South Bay. This list begins at the eastern end of the San Mateo Bridge and proceeds clockwise.

From the last exit (Clawiter Rd./Eden Landing Rd.) on Highway 92 before the bridge, you can easily reach the Hayward Regional Shoreline Interpretive Center on the north of the highway. Walking north from this trailhead takes you along a stretch of the original shoreline to the site of John Johnson's landing. To the south of the highway lies the trailhead for the Eden Landing Ecological Reserve. Following the Bay Trail south from the parking area along the edge of the reserve leads

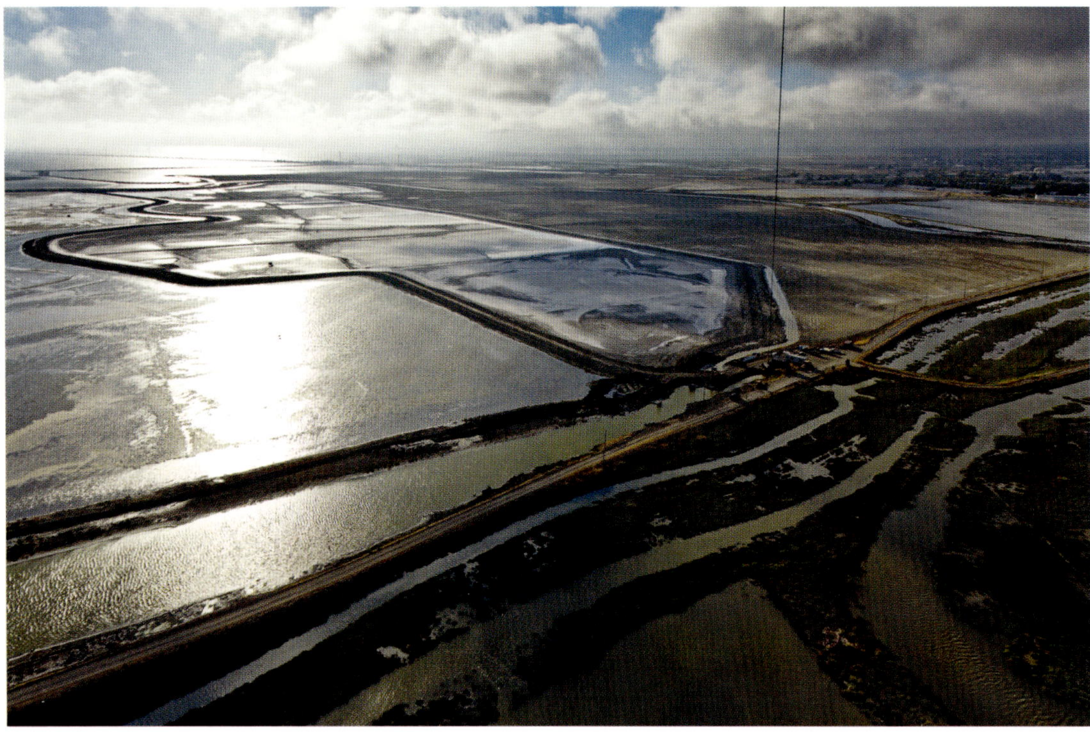

The Eden Landing Ecological Reserve from above Mount Eden Creek Marsh. As of 2013, public access to the reserve is limited to the section of Bay Trail traversing its eastern acreage. Plans are underway for public access to the Oliver Salt Works ruins.

to the recently established Mount Eden Creek Muted Marsh. A modest hike further down the trail brings you to the curious Eden Shores housing development, a recent herniation of development into the South Bay flats that features a moat to deter house cats from entering the reserve.

Further south, near the Dumbarton Bridge approach, are many places to explore via hikes staged from the main parking areas of the Don Edwards San Francisco Bay National Wildlife Refuge off Thornton Avenue in Newark. To the north you can cross Highway 84 using the pedestrian bridge atop the toll plaza. This leads to the Coyote Hills, with miles of hiking through interesting topography: salt ponds to the west and tule marsh to the east. Note the Red Hill Gravel Quarry as you exit the pedestrian bridge. This quarry, a 150-foot-tall hill in 1974, was excavated to a depth of 320 feet below sea level before closing in 2010. To the east of the Don Edwards parking area you find the LaRiviere Marsh Restoration Project. Once the site of Arden Salt Works #1 and Jarvis Landing, it now sports a returning mantle of marsh vegetation and associated wildlife. To the west lies Salt Pond N1, the pond I encountered in my first visit to the South Bay. You can circumambulate this pond in a pleasant five-mile walk that takes you along the free-flowing Newark Slough, past the Dumbarton Cutoff Line swing bridge, and along stretches of Dumbarton Marsh. A short drive from the Don Edwards parking area takes you out along the old 1924 highway to Dumbarton Point. There a pleasant fifteen-minute walk south along the bay-shore levee brings you close to the Hetch Hetchy pipeline crossing and the Dumbarton Cutoff Line. Heading north under the main highway bridge connects you with long and interesting trails through the salt ponds and past a section of original bay shore and marsh. You can follow these levees to the north end of the Coyote Hills.

Heading further south from Don Edwards by car affords the opportunity to drive by the Cargill Salt plant and its huge stacks of salt. The parking lot contains an informal collection of old salt-harvesting machinery and a reproduction of the Oliver-designed, wind-driven Archimedes screw pumps. Continuing southward on Cushing Parkway past the Fremont Auto Mall leads to the trailhead for the Warm Springs Seasonal Wetland, where a short hike provides glimpses of the Warm Springs vernal pool habitat and the adjacent Salt Pond A22, with its dugout duck blinds.

The south end of the bay offers Alviso, a fun, eclectic town to visit, and opportunities for several fine hikes. The Don Edwards Environmental Education Center in Alviso, off Zanker/Los Esteros Road, lies nestled between landfills, the impressively benign outflow of the Santa Clara Pollution Control Plant, and the New Chicago Marsh. The Center is worth a visit for its programs and friendly staff as well as a boardwalk that takes you through a small section of the New Chicago Marsh. The center is also

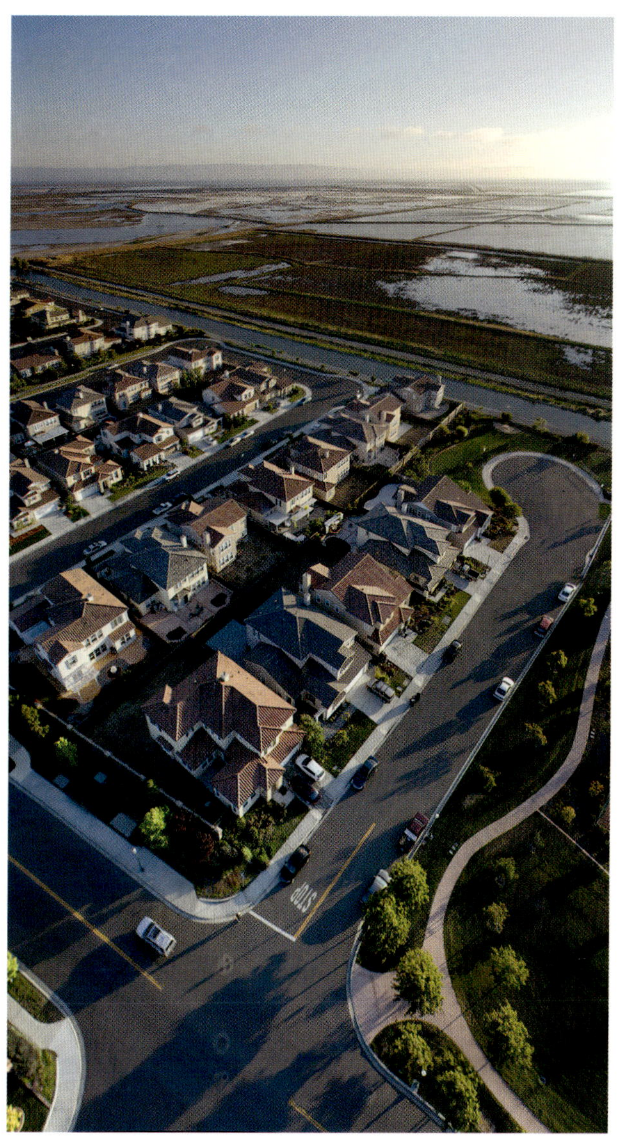

The Eden Shores housing development (Hayward, July 2010)

a trailhead for longer hikes, around Salt Pond A16 and to points beyond. A short drive or bicycle ride away, the Alviso Marina County Park offers a staging area for long hikes north into the salt ponds. These can take you along the old, and still active, 1880 railroad grade, past sections of original marsh and salt ponds nearing retirement, and up to the closest public viewing point for the abandoned town of Drawbridge. It is fun to poke around Alviso, once the port for San Jose, and include a walk down the levees containing Alviso Slough, levees that are quite critical due to Alviso's substantial subsidence.

Wrapping around to Ravenswood on the west shore of the bay, Salt Pond SF2 is worth a visit. Lying between the contemporary Dumbarton Bridge and the Dumbarton Cutoff Line to its south, Salt Pond SF2 is an experiment in habitat creation with advanced flow control structures managing tidal flow, a collection of compact islands for nesting, and an area set aside as snowy plover habitat. A short hike at SF2 takes you along the marshy bay shoreline past two newly constructed wildlife viewing platforms.

Further north is Bayfront Park in Menlo Park, a plot of land that hosted marsh, salt ponds,

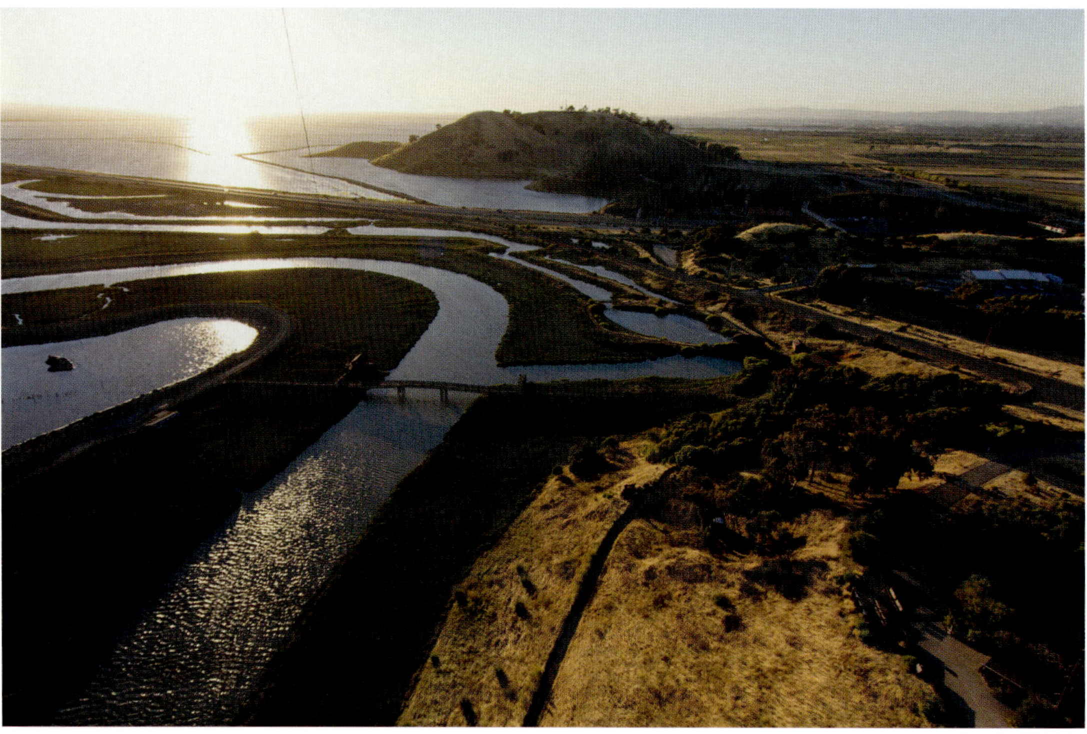

The Coyote Hills are a splendid place to hike. The higher points, at 300 feet above sea level, offer views across the salt ponds and adjacent flatlands. The trails at sea level wander past salt ponds and tule marshes.

and a municipal landfill before its current incarnation as publicly accessible open space. Bayfront Park is notable for its neighboring Salt Ponds R4 and R5, disused ponds that tend to sport the bright colors of high salinity. This is your best bet for a close encounter with extreme halophiles. From the hills of Bayfront Park you can gaze westward to see the crystallizer beds of Cargill Salt's Redwood City plant, a site that can also be viewed from Redwood City's Seaport Boulevard.

Books and Reports

I have had a fine time reading, on a nice rainy day or in anticipation of a hike, background information relevant to the South Bay landscape. My favorite book to date has been William E. Ver Planck's *Salt in California*, published in 1957 as Bulletin 175 for the Division of Mines in the State of California's Department of Natural Resources. This tidy, precise document offers insightful history, descriptions of industrial process, maps, and more. A great companion book for *Salt in California* is Harold Gilliam's *San Francisco Bay*, published in the same year. Therein Gilliam offers an affectionate portrait of the bay, with sections specific to the South Bay, from a temporal vantage point right on the threshold of major change. Like all of Gilliam's books, it is a fun read. John S. Sandoval's *Mt. Eden: Cradle of the Salt Industry in California* (1988) provides a useful accounting of the salt industry's development and subsequent maturation in the Mount Eden area. Sandoval offers glimpses into the lives of early salt entrepreneurs and the communities they formed. *Drawbridge, California: A Hand-Me-Down History*, O. L. Dewey's slim 1989 volume, provides a similar glimpse of life a bit further south in the bay where the hamlet of Drawbridge developed, prospered, and waned. Matthew Booker's *Down by the Bay: San Francisco's History Between the Tides* offers an enjoyable accounting of the use and development of San Francisco Bay tidelands and includes an insightful section on the South Bay.

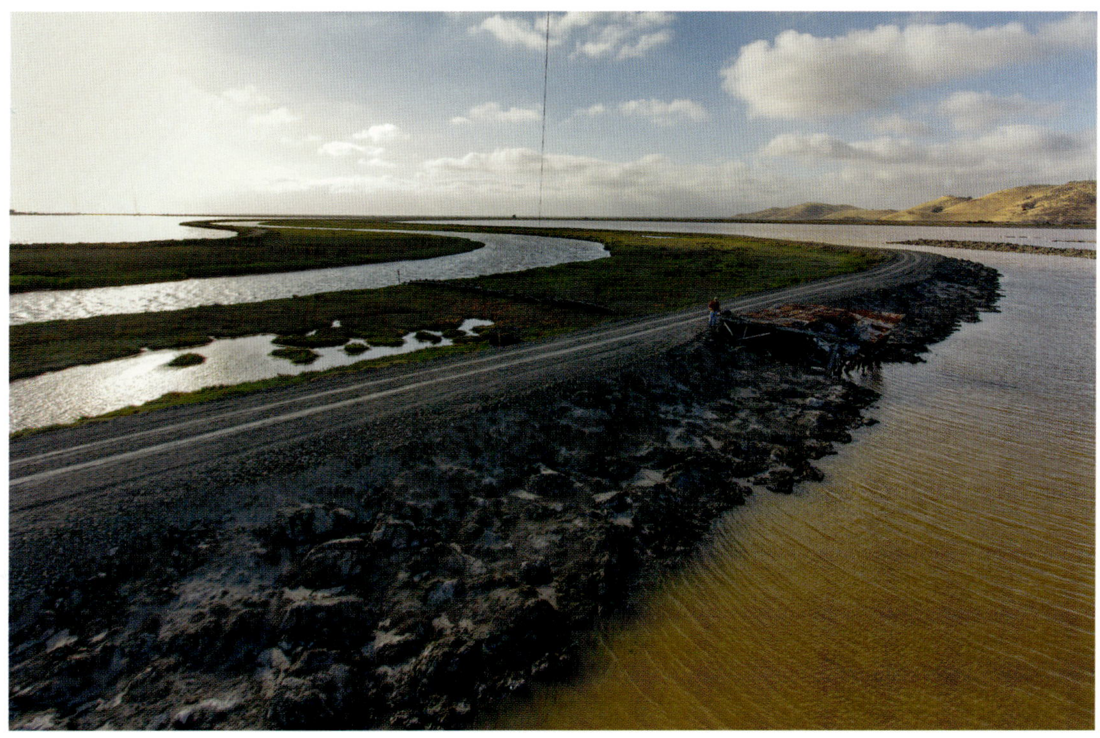

Along the Newark Slough Trail Loop, adjacent to Salt Pond N1, you will see the marsh fringes of Newark Slough, a large strip of the Dumbarton marsh, and the Dumbarton Cutoff Line swing bridge. It is an afternoon well spent.

Mark Kurlansky's book *Salt: A World History* (2002) is useful for placing the South Bay salt industry in the larger context of salt and culture and it is a fun read as well. The elegant *San Francisco Bay: Portrait of an Estuary*, a 2003 collaboration between John Hart and photographer David Sanger, parallels Gilliam's book in that it provides a summary of bay issues right at the threshold of major change. Like Gilliam's, Hart's text provides a fun and insightful read plus there is the considerable bonus of Sanger's evocative photographs. An interesting complement to John Hart's text is *Natural History of San Francisco Bay* (2011) by Ariel Rubissow Okamoto and Kathleen M. Wong, which examines the bay's environmental challenges in greater detail from a vantage point one decade later. The Center for Land Use Interpretation's *Back to the Bay: Exploring the Margins of the San Francisco Bay Region* (2001) frames a more prosaic view of the bay's edge, with attention to past uses that have shaped the shoreline. Matthew Coolidge, author of *Back to the Bay*, has a knack for focusing attention on everyday types of land use so commonplace they disappear in plain sight.

A number of interesting reports emerged from feasibility planning in preparation for the 2003 grand bargain that began the South Bay Salt Pond Restoration Project. Many of these can be found online. The centerpiece is certainly the *Baylands Ecosystem Habitat Goals* report (1999) prepared by a large team of researchers pulled from many public and private agencies. The report is substantial, providing both technical detail and a fascinating window into the reasoning of the day and the formation of an ambitious project. In 2002 S. W. Siegel and P. A. M. Bachand issued the comprehensive *Feasibility Analysis of South Bay Salt Pond Restoration*, a detailed, pre-purchase assessment of many nitty-gritty aspects of South Bay wetland restoration. Save the Bay produced the more compact *Turning Salt into Environmental Gold*, a 2002 assessment that provides a relatively concise summary of restoration potentials and challenges. As the restoration project gathered steam the Bay

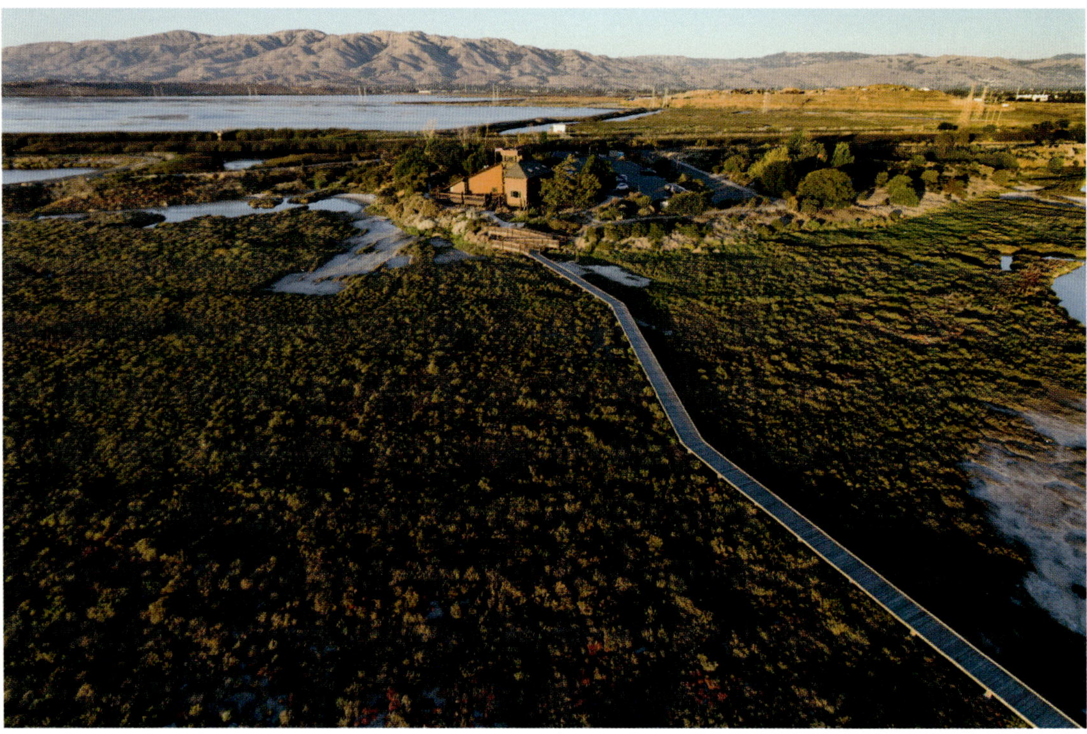

The Environmental Education Center, adjacent to the New Chicago Marsh and part of the Don Edwards San Francisco Bay National Wildlife Refuge, provides exhibits and classes interpreting the South Bay landscape. It is a fine starting point for a South Bay hike.

Conservation and Development Commission issued a staff report, *Salt Ponds* (2005), that provides a great summary of restoration progress and potentials as well as a declaration of policy at that point. Also in 2005, the consulting firm EDAW issued a *Historical Context Report* for the restoration project, a document containing useful background and observations. This was followed in 2009 by the US Fish and Wildlife Service's report *Identification and Evaluation of the South San Francisco Bay Solar Salt Industry Landscape*, which suggests that the cultural geography aspects of the project are somewhat adrift.

Maps and Aerial Survey Photos

Having long had cartographic proclivities, I have experienced many entertaining hours poring over old maps of the South Bay. I began with the tactile pleasure of handling large sheets in UC Berkeley's fine Earth Sciences and Map Library. In the current day many of my favorite maps are available online in high resolution.

A great first stop for South Bay cartographic resources is the website of the Historical Ecology group at the San Francisco Estuary Institute (http://www.sfei.org/he/HE-publications). Here Robin Grossinger's fine team has assembled a first class set of resources. For instance, their "T-Sheet User Guide" will lead you to high-resolution versions of all of the 1850s US Coastal Survey charts of South San Francisco Bay and provide an interesting accounting of the history and methods of that survey. The group has been prolific in developing insightful assessments of the bay environment and its history. Spend some time reviewing their products, such as the EcoAtlas Baylands Maps or the Creek and Watershed map series (in conjunction with the Oakland Museum of California) and you will find it time well spent.

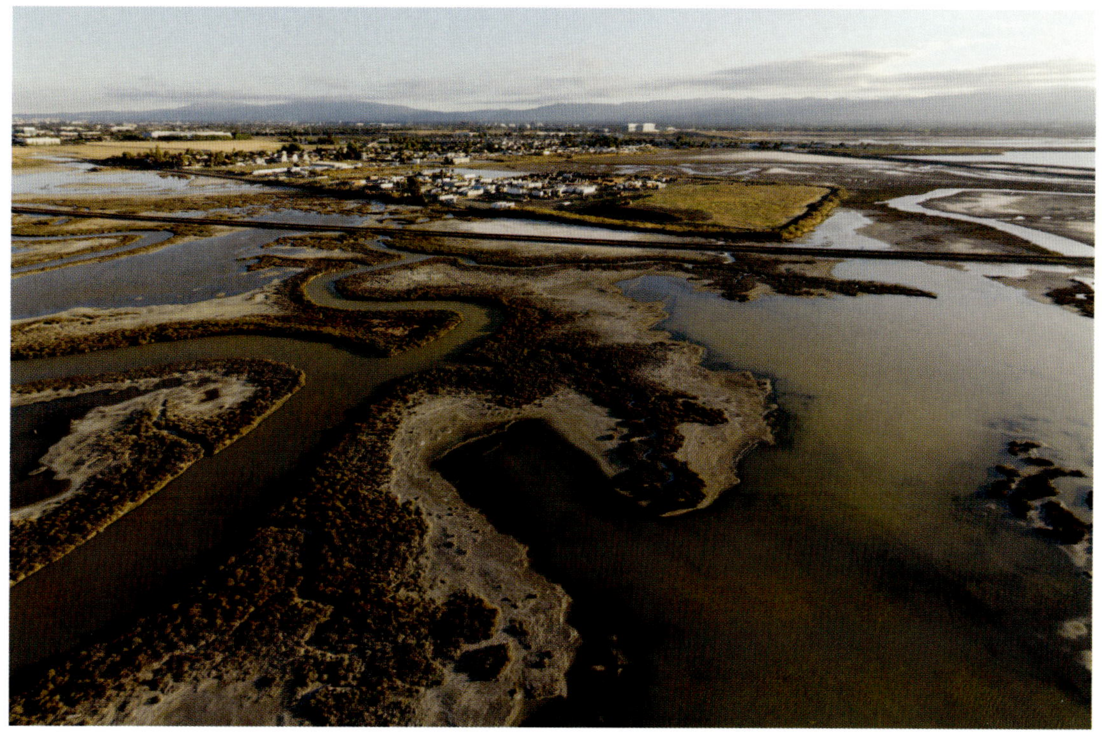

Alviso, as seen from the New Chicago Marsh. The city of Alviso was once the port for San Jose. Trails here wander along with the Guadalupe River and out into the salt ponds.

Another Internet resource is the David Rumsey Map Collection (www.davidrumsey.com). Rumsey, a dedicated collector of old maps, has made his collection available in high resolution online. Of particular interest for the South Bay are Rumsey's reproductions of the 1870s Thompson & West Historical Atlas Maps for Alameda and Santa Clara counties, with maps depicting properties and land use as well as etchings of old saltworks and boat landings.

The South Bay Salt Pond Restoration Project website (http://www.southbayrestoration.org) map section includes thematic maps for the project, including the definitive map for salt-pond numbering nomenclature. Also on the site is an interesting interactive map feature where, with patience, you can pull up aerial photographs of the project areas dating from 1928!

With a bit of searching you can find old USGS cartographic quadrangles on the Web. Coverage of the South Bay goes back to 1899 and the old versions provide a useful view of salt-pond development. And of course there is Google Earth, an online resource that is extraordinary by any measure.

NOTES TO PLATES

1.
The red tones of higher salinity ponds become rich and subtle in the warm light of dusk. It is a time of day when the colors blend and lines soften to good effect.

2.
In one of my early photographs of salt-pond shorelines, a residual ring of salt formed as the pond's brine level receded due to evaporation. Here and there small weeps appeared where groundwater drained through the salt crust to leave a mark, as is evident in the upper right of the image.

3.
A corridor on the Dumbarton Point peninsula holding the Hetch Hetchy water pipeline and the Dumbarton Cutoff rail line also hosts a constrained strip of tidal marsh. Near what was once Plummers Landing, this stretch of marsh is sandwiched between Cargill's bittern ponds (blue in the image) and Salt Pond DP1 near the Don Edwards Refuge headquarters.

4.
This crooked levee separating Salt Pond 13 from Salt Pond A15 follows what was once the edge of Grey Goose Slough. The slough was a major marsh channel that was first interrupted by the construction of the 1880 South Pacific Coast Railroad and then eliminated altogether by the construction of twentieth-century salt ponds. Its shape, however, remains in the levee.

5.
This self-portrait in repose was taken on the dry surface of an old duck pond situated on the leeward side of the Coyote Hills. These ponds fill with rainwater seasonally and are typically dry by late summer.

6.
Each little section of shoreline is a marvel for the smooth gradients of color presented as pond depth increases and for various manifestations of salt crust and residue.

7.
This salt evaporation pond near Mowry Slough is approaching the salinity required to precipitate salt, a circumstance that produces rather startling colors associated with archaebacteria.

8.
Each salt-evaporation pond has its own character, with sublte variations in color and surface tension that cycle with season, weather, and orchestrations of the salt makers.

9.
The colors and textures of inactive ponds reflect more dramatic environmental conditions from their past. This pond bottom retains salt, and associated microorganisms, from its days as an active evaporator in the twentieth century.

10.
This shallow mud levee runs between a distribution channel containing bay water and a former salt pond loaded with a substantial amount of residual salt. As the summer season progresses the rains of winter evaporate from inactive ponds, salinity increases, and different halophiles prosper.

11.
This bleached white skeleton is a former bush that died when a small salt pond at the north end of the Coyote Hills was connected to tidal flow via a culvert. The crackled ground texture surrounding the skeleton is a harbinger of greater change associated with the new tidal regime.

12.
As salinities vary from pond to pond, so does color, with each shade revealing something about the salinity of a pond and the community of halophiles it supports.

13.
This shoreline detail shows fresh levee only a few weeks old against the turbid green brine of a medium-salinity pond.

14.
The low point in a relatively flat drainage system, this ditch bottom along the railroad right-of-way leading into Alviso cycles through an interesting range of colors under the influence of tides, season, and the state of adjacent salt ponds.

15.
This view displays the warm colors of an inactive salt pond as it fills with winter rainwater. The Coyote Hills are visible in the distance.

16.
Many pond bottoms evidence a curious buckled texture that I have come to associate with gypsum (calcium sulfate), a mineral that precipitates from medium-salinity brine.

17.
There are tire tracks aplenty in this close view of a recently harvested crystallizer bed near Cargill's Newark plant. Having removed the bed's annual yield of salt, equipment was hard at work to level the bed for the next year's cycle.

18.
This view shows two crystallizer beds during harvest season. The upper bed has been drained and displays the tracks left by the harvesting machinery as it gathered salt. The lower bed awaits harvest with a load of magenta "pickle" that will soon be drained.

19.
This scene contrasts the yellow-orange tones of an inactive salt pond undergoing remediation with the cloudy greens of the recently restored North Creek Marsh tidal flats.

20.
Salt ponds are often joined by a dredge lock—a small split in the levee that allows a dredge to make a transition from one pond to the next with minimal exchange of brine. These two ponds in the Mowry complex color their dredge lock with the bright hues of halophiles. A small strip of marsh is also visible.

21.

This lone bush has taken root near the borrow ditch of Salt Pond R5, an inactive pond that changes conditions with the season. The bush is one of several along the pond's edge as it meets the promenade of Bayfront Park in Menlo Park—one of the best places to view highly saline pond conditions close up.

22.

The former course of Beard's Creek is now a long, linear salt pond filled with turbid green water. This old flow control gate has survived longer than the small levee that contained it.

23.

Traces of former marsh channels abound at Bayfront Park.

24.

The salt ponds near Bayfront Park in Menlo Park have been out of production for a decade. Over this time their exposed, dry plateau has developed a light crop of salt-tolerant plants.

25.

During the summer Salt Pond E14 often goes dry. In this view it displays the remains of an unquestionably sinuous marsh channel.

26.

Salt Pond E8X is a small, curious pond that has changed character repeatedly during the years I have been photographing the area. This view dates back to a period before it was connected to the tides.

27.

The high-salinity evaporation ponds flanking Mowry Slough present a striking red color. As the cycle of evaporation progresses, the plateau of the pond begins to emerge with details of small channels and buckled gypsum. The pond's borrow ditch is visible in the upper right.

28.

In the masking glint of a setting sun it is easy to imagine New Chicago Marsh in its natural state. It has the visual patterns of an active marsh but lacks tidal ebb and flow in our day.

29.

In 2008 a large forest fire near Big Sur kept the Bay Area skies hazy. For a week or so the sky would glow particularly red in the diffuse light of sunset to complement the Mowry pond landscape, which was also red. It was a striking visual effect.

30.

The higher edges of the larger channels in this view are exposed, while the pond's plateau remains underwater. This network of marsh channels is interrupted by the pond's borrow ditch, which spans the top of the image. Interruptions of this sort are a challenge when preparing a salt pond for restoration of tidal flow.

31.

The medium-salinity ponds near Alviso often display a bright, almost fluorescent, green color. In this image it contrasts with the dark gray of exposed mud in a scene that features familiar marsh textures.

32.

The plateau of Salt Pond R5 holds a distinct pattern of former marsh channels. This pond has not been flooded in years and has developed a surface texture that combines a fine-grained, soft buckling with small tufts of vegetation.

33.
This large marsh channel lies trapped between the Alviso Flat and a local landfill.

34.
As I found marsh channels in the aerial views of various salt ponds I was impressed at how widespread the channel network had been and by the rich variation in how the channels manifest from pond to pond. Each pond, it seemed, had a unique character in tonality and texture.

35.
The salt-evaporation pond bottoms are interesting for their variety. This active, medium-salinity pond features soft, wispy traces of former marsh channels and a complementary palette of subtle colors.

36.
The pond bottoms near the old Oliver Salt Works are filled with the traces of even earlier saltworks. This eroded fence line might date back more than a century.

37.
This decaying bridge over what appears to be a small salt pond was instrumental in piquing my early curiosity about the South Bay landscape. It turns out this was the site of the Union City Salt Works and the bridge crossed what was once Alameda Creek. The creek has now been shifted to a flood control channel.

38.
A tight swarm of shorebirds flies over an old levee near the Oliver Salt Works ruins. The birds coalesce for protection from predators and fly with a fascinating precision.

39.
This landing, located near the eastern anchorage of the current-day Dumbarton Bridge, is the most detailed landing ruin in the South Bay. The docks are located on what was once the channel of Beard's Creek and were apparently connected to a railroad spur.

40.
Long rows of piles line what appear to have been covered basins. The roof structure is long gone and the pools alternate between dry and flooded, depending on the season.

41.
The Oliver Salt Works ruins contain a densely packed collection of foundations, tanks, and infrastructure. This image shows the end of a long series of below-grade troughs of unknown function.

42.
As in many of the large contemporary salt ponds, there are traces of an earlier saltworks in the bottom of Salt Pond E8. The location vaguely aligns with the Occidental Salt Works on Ver Planck's map.

43.
Salt Pond E8X is bisected by the low traces of early saltworks structures. In this image the faint outlines of these elements are visible within a large mat of cyanobacteria.

44.
Once an early brine distribution system, the straight line running diagonally across this image leads to the Oliver Salt Works plant site (visible at the top of the image). The brine crossed a large borrow ditch on a raised trough.

45.
Small levees subdividing the bottom of Salt Pond E13 date back to earlier saltworks and reveal a network of distribution ditches and raised troughs used to move brine from pond to pond.

46.

This example is typical of many of the nineteenth-century traces. Unnoticed from the ground, regular shapes emerge in an overhead view from features that are only a few inches high. In this case they are the rectangular outlines of small impoundments or structures.

47.

In this overhead view of a derelict Archimedes screw pump, the intake ditch has silted to a shallow round shape. Follow the longitudinal axis of the pump and you can see the screw structure where it enters the intake ditch.

48.

The Oliver Salt Works site was once covered by multi-story plant buildings and holding tanks. In the center of this image you can see the round foundations and collapsed compression straps as all that remains of large wooden storage tanks.

49.

This detail of the Oliver Salt Works ruins shows a regular array of foundation pilings in a crawl space that extended below grade. Snowy plovers often nest adjacent to this site.

50.

The Oliver Salt Works subsumed the nineteenth-century Whisby works. This view west of the Oliver plant probably contains traces of both. The raised raceway once carried salt brine from pond to pond.

51.

This old paddle-wheel pump appears to be an early, site-built electric pump. Three power poles lead to its location but the motor is long gone. The pump currently serves as the base for a large heron's nest.

52.

The abandoned dwellings of Drawbridge flank the Union Pacific rail line and are in turn flanked by Salt Ponds A20 and A21. The twentieth-century expansion of salt ponds reduced the acreage of natural marsh surrounding Drawbridge to a small fraction of its former size. This led to a decline in the natural amenities that once underpinned the hamlet's popularity.

53.

This is a southward view toward Mowry Avenue with the rail corridor separating crystallizer beds on the right from office park developments on the left.

54.

In this scene the outermost bodies of brine are inactive salt ponds (A13 left and A16 right), then immediately flanking the straight Union Pacific railroad grade are the Alviso Flat on the left and New Chicago Marsh on the right as they taper into ditches. The scene is animated by colors related to the salinity of each body of brine and its resulting communities of microorganisms—très jolie.

55.

This Drawbridge dwelling has nary a roof shingle as nature conducts its slow deconstruction of the building. The rail bridge in the distance spans Coyote Creek.

56.

This overhead view shows a fast-moving Amtrak train as it zips down the Union Pacific rail corridor. Amtrak rolls through regularly, and at a consistently brisk pace.

57.

The Newark Slough bridge sits within one of the larger remaining areas of marshland on Dumbarton Point. The slough, still free flowing, is navigable up to the former location of Jarvis Landing at the southern end of the Coyote Hills.

58.
The Newark Slough swing bridge is a marvelous old structure. Built of standard steel sections with hot rivet connections, the bridge is an elegantly functional artifact from the golden age of railroads.

59.
The blossoms of invasive perennial pepperweed surround wooden structures in decline at Drawbridge. Pepperweed seems to prosper in the South Bay. Drawbridge's two organizing elements—the railroad grade and a navigable marsh channel—are visible in the image.

60.
This view of the north end of Drawbridge looking toward the south illustrates the tight siting of dwellings between the Union Pacific rail line and what was once a navigable marsh channel.

61.
The ditches flanking the Union Pacific rail line accrue salt through windblown spray and foam. As ditch water comes and goes through the seasons its salinity changes, microorganism populations flourish and decline, and colors vary. This makes for an entertaining ditch.

62.
A view toward the south showing what remains of Drawbridge's small collection of structures. Where once there were a hundred or so structures only a few dozen remain. These are returning to the earth quite rapidly, sometimes suddenly in fires set by vandals.

63.
In the south end of Salt Pond E6B, dugout duck blinds have been built using a rectangular wood box construction. As with other dugout blinds, a small hummock was created to keep the pond waters at bay.

64.
The eastern edge of Cargill's Newark crystallizer beds presents a tidy, compressed edge between the saltworks and Newark. Compactly aligned in this image, from left to right, are the crystallizer bed, a levee with road surface, a concrete-lined streambed, and railroad tracks.

65.
This view of a recently refurbished levee illustrates one form of levee erosion in action. In winds peaking over 20 mph, an almost continuous stream of windblown dust, visible in the image, transports matter from the levee to the adjacent ponds. Over time the levee inevitably loses height and integrity.

66.
These old, irregular levees created small ponds. I originally thought they were late-stage ponds in one of the early saltworks but Cargill staff familiar with the area explained they created freshwater ponds for duck hunting.

67.
At the Eden Landing Ecological Reserve new levees constructed with imported fill present a different texture than the mud clumps of the dredged levee construction traditional to the area.

68.
While some South Bay levees follow the outline of natural features in the landscape, others are the product of land subdivision. These levees follow lines that originated on paper.

69.
This low-level aerial image shows the *Mallard II* clamshell dredge as it renews the levee along the former course of Beard's Creek. In the image you can see a second layer of dredged mud being placed by the dredge along the centerline of the levee.

70.

This view shows the plateau of Salt Pond A21 a couple of years after it was returned to the tides. The buckled gypsum of the plateau is beginning to soften with deposited sediment while the borrow ditch is acquiring the patina of a biological film. The sharp edges of the borrow ditch and its dredge marks were quick to soften in the years after this photograph.

71.

This two-toned borrow ditch appears to have a bottom profile with a deep center section and shallow sides. Or perhaps the visual effect is due to the deepest portion hosting a different halophile with darker tones. Dredge marks are evident along the edge of the plateau.

72.

Even in ponds long dry, marks of dredging are easily recognized. These divots in Salt Pond SF2 were sharply defined before they were erased altogether by the pond's reconfiguration.

73.

This wonderfully graphic line of dredge marks is offset from the original borrow ditch. Their scale as seen in the photograph indicates a relatively high camera position.

74.

Salt Pond A23, an inactive pond, fills with the winter rains and evaporates to near dry in the summer. It seems to have a high residual salt content, for the pond always displays interesting colors. It consistently offers photogenic scenes such as these fading dredge marks.

75.

These secondary dredge marks pepper a stretch of Eden Landing levee. The marks sit below a shallow layer of pond water and were not visible from the vantage point of the levee. Finding them in the photograph was a complete and welcome surprise.

76.

This collection of soft and irregular dredge marks occurs at the corner of a salt pond. At corners the radius of the levee is larger than that of the ditch, thus requiring extra scoops for levee renewal.

77.

The sharp dredge marks in this scene look rather fresh but the ditch itself seems shallow and soft. I suspect the marks are fairly old.

78.

This old borrow ditch in an inactive pond has almost gone dry. A salt crust has formed over what's left of the winter rains, and faint round divots serve as testimony to dredge scoops from long ago.

79.

These soft dredge marks have become round with age and associated erosion. The marks cut off from the borrow ditch have dried completely to leave a small crust of white salt.

80.

This view shows Salt Pond E8 sporting a motif of earth colors and features a jagged row of dredge marks in the inside edge of the pond's borrow ditch. Their diminutive appearance suggests the photograph was taken from a relatively high vantage point.

81.

This January view shows a raised duck blind in Salt Pond E14 under winter hunting-season conditions. The pond is flooded to a shallow depth with a combination of rainwater and water from the bay. Salt Pond E9 and the Coyote Hills are visible in the distance.

82.

Once one of the more elaborate duck hunters' cabins, this abandoned structure in Salt Pond E6B is now the site of a half-dozen herons' nests. The building has split open to reveal interesting glimpses of its interior.

83.

Like the dredge marks in the lower portion of this image, duck blinds are scattered across much of the South Bay landscape. This example, a duck blind visible in the upper left, seems to be a simple wooden box.

84.

This painterly collection of footprints crosses the orange shoreline of Salt Pond A23 to tell the story of predawn embarkation on short boat rides to various duck blinds. The scene includes a beer can as a scale clue.

85.

The raised duck blind in the upper portion of this image lends scale to the large mudflat near the mouth of Mount Eden Creek. Once the site of a nineteenth-century saltworks, a trace of which is visible in the top of the image, this mudflat is reestablishing itself as a natural feature.

86.

For much of the year the raised, shacklike duck blinds of our day stand as silent and neglected sentinels waiting for the short wintertime hunting season.

87.

The large, raised wood blind in the foreground of this image reveals yet another type of blind construction in Salt Pond E14. These raised blinds would seem to offer greater amenity than the early dugout blinds and are certainly more spacious. This particular blind contains three chairs.

88.

The Environmental Education Center in Alviso offers a very pleasant boardwalk that takes visitors on a small loop through New Chicago Marsh. The path meanders past colorful marsh channels, pannes, and a borrow ditch.

89.

This image captures the narrow and tidy dike construction that defines Cargill Salt's array of crystallizer beds. The pile of salt in the center is the remains of a salt ramp constructed to allow equipment to drive from bed to bed. Also visible are remnants of the narrow gauge railroad track from the old system used to transport the salt harvest.

90.

Equipment tracks and colors combine to create an abstract pattern in a drained crystallizer bed. It is an interesting and ephemeral record of a harvest recently completed.

91.

This view of the salt harvest in progress shows a tractor-pulled conveyor lifting previously ground salt into the bed of a balloon-tired dump truck. The truck slowly parallels the conveyor and is filled with thirty-five tons of salt in about ninety seconds.

92.

This graphic view includes a harvested crystallizer bed at the top of the image complete with parallel swaths left by the harvesting equipment. Horizontal lines at mid-image are a salt-crusted levee, the old narrow gauge rail track, a levee-top roadway, a brine distribution ditch, and a vegetation-covered levee.

93.
This view to the southwest across Cargill Salt's tidy Newark landscape includes a recently harvested crystallizer bed at the bottom, a bed awaiting harvest on the right, and a holding pond on the left.

94.
Cargill Salt's storage stacks are smaller at the beginning of the fall harvest season and grow quickly as the harvest progresses. This view, taken from out in the crystallizer ponds, features the Coyote Hills in the distance.

95.
Each year's salt harvest is gathered at Cargill Salt's Newark plant for storage and subsequent processing. The harvested salt is transported by conveyor belt and gantry to form twin stacks around five hundred feet long and several stories high. Bulldozers work to shape the top of the stacks and later remove salt for processing.

96.
In this photograph the sun glints across the drained bottom of the largest crystallizer bed in Cargill Salt's Newark complex. Mowry Slough is visible in the distance with open bay waters beyond.

97.
This tractor pulls laser-guided leveling equipment to prepare a recently harvested crystallizer bed for the next annual cycle of salt production. The bed will be filled again with highly concentrated brine (pickle), which will precipitate sodium chloride to form a new layer of harvestable salt.

98.
Each band in this richly detailed image records the repeated passes of the salt-harvesting machinery as salt was loosened, gathered, and finally loaded onto dump trucks for transport to the washhouse.

99.
Salt Pond SF2 at the western anchorage of the Dumbarton Bridge was dry until recent years. This image shows its state during the summer, when the pond bottom served as nesting ground for snowy plovers. A portion of the pond will be maintained in this state for that purpose.

100.
Salt Pond SF2, once a barren gypsum plain near the western anchorage of the Dumbarton Bridge, was redesigned as an experiment in avian habitat. The new pond layout includes an advanced tidal flow control system and a series of small islands.

101.
Salt Pond A6, also known as the Knapp Tract, was recently home to thousands of nesting California Gulls. The gulls were displaced when the pond was returned to tidal flow in December 2010. Before the levee breach, the pond bottom featured a large network of dormant marsh channels as shown in this photograph.

102.
Salt Pond A6, shown in the upper portion of this photograph, was quick to accumulate sediment after its levees were breached. Within months the pond's old bottom textures were erased and replaced by the smooth sediment of a mudflat.

103.
The last few years have seen the installation of many new flow control gates in preparation for renewed tidal flow across large portions of the Eden Landing Ecological Reserve. This new gate will help control the salinity of Salt Pond E13 as it enters a new management regime focused on habitat creation.

104.
The Don Edwards San Francisco Bay National Wildlife Refuge now controls large swaths of the southernmost reaches of the bay, including New Chicago Marsh. This view over the marsh toward Alviso shows a landscape that features remnants of tidal channels trapped between levees and the uplands.

105.
This new flow control structure was activated in 2012 to allow tidal flow from Mount Eden Creek into Salt Pond E12. The restoration project calls for a revised set of ponds with varied salinities to stretch across the area once occupied by Salt Ponds E12 and E13.

106.
The bay edge of Salt Pond SF2 now features an improved promenade with two raised viewing structures. From these vantage points you can observe local wildlife and the pond's rather intricate system for managing tidal flow.

107.
The vernal pools of the Warm Springs Seasonal Wetland develop interesting concentric rings of flowers as they dry out in the spring. Each species establishes its place according to the water depth it can tolerate. As the pool evaporates, the flowers present a record of its wet-season depth.

108.
In preparation for returning Salt Pond A6 to tidal flow, restoration crews constructed a strategic set of ditches to reconnect marsh channels previously interrupted by old levees. This process leveraged lessons learned in previous breaches to prepare the pond for effective tidal flow.

109.
The North Marsh area of the Eden Landing Ecological Reserve was connected to tidal flow in 2006. Since that time ground textures have changed and marsh vegetation is taking root. This curved levee is at the northern end of the marsh.

110.
This view of Mud Slough, once known as Warm Springs Slough, shows Salt Pond 23 to the left and Salt Pond A20 to the right. Mission Peak is visible on the horizon. Late afternoon is a pleasant time in the South Bay.

ACKNOWLEDGMENTS

My kite aerial photographs, the wanderings they engendered, and ultimately this book would not have been possible without the generous encouragement and support of many along the way.

My early attempts at lofting cameras with kites would surely have failed without advice kindly offered by Brooks Leffler, Anne Rock, and Craig Wilson. Kite aerial photographs by George Lawrence and Arthur Batut, taken more than a century ago, served as inspiration as have the contemporary photographs of Pierre Lesage.

Mark Frauenfelder and Dale Dougherty at *MAKE* magazine shared enthusiasm for my tinkering at just the right moment to sustain and advance the work.

I owe a great debt to the Exploratorium for providing safe harbor during a 2003 sabbatical leave. This marked the beginning of the Hidden Ecologies Project and work on the South San Francisco Bay salt-pond landscape. In particular, Senior Artists Peter Richards and Susan Schwartzenberg at the Exploratorium provided encouragement and an opportunity to refine the work. Marina McDougall contributed effectively to the early framing of the project while Wayne Lanier, companion muse on many a hike, has been a steadfast and valued collaborator.

Staff at the Don Edwards San Francisco Bay National Wildlife Refuge, managed by the US Fish and Wildlife Service, and the Eden Landing Ecological Reserve, managed by the California Department of Fish and Wildlife, provided patient counsel and, importantly, special use permits to take kite aerial photographs in the refuge. Gaining permission to fly kites for aerial photography in a wildlife refuge would not have been possible without managers blessed with imagination and initiative. Clyde Morris and John Bradley of the Don Edwards Wildlife Refuge were instrumental in granting early permissions. Multi-agency, multi-year access followed through the generous stewardship of Steve Richie, then director of the South Bay Salt Pond Restoration Project at the California Coastal Commission. John Bourgeois, the project's current director, continued this stewardship with nary a hitch. My day-to-day explorations of the South Bay have been enriched by the extensive practical knowledge and consistent good cheer of wildlife biologists Cheryl Strong and John Krause. Jennifer Heroux, Debra King, and Ivette Loredo have lent support along the way. Mary Selkirk and Will Travis have been effective advocates for the dissemination of my South Bay work.

I am most grateful to Cargill Salt for offering photographic access to their privately held South Bay works. Jill Singleton skillfully negotiated this arrangement and Terry Lewis served as a knowledgeable and entertaining field guide. Cargill's Barbara Ransom later provided enthusiastic and welcome support for this book project.

My UC Berkeley colleagues provided patient counsel and critique during the course of this project. I am particularly grateful to Nezar AlSayyad, Ed Arens, Gary Brown, Paul Groth, Susan Ubbelohde, Linda Vida, and Jennifer Wolch for their encouragement. The UC Berkeley Committee on Research provided modest support for project expenses.

I am grateful to Malcolm Margolin, Gayle Wattawa, and the fine staff at Heyday for imagining the potentials of this book and making its realization a pleasure.

And my greatest gratitude goes to my family, Claudia Benton and our sons Charlie and Thomas, for their companionship and support as kites and the South Bay became part of our family life.

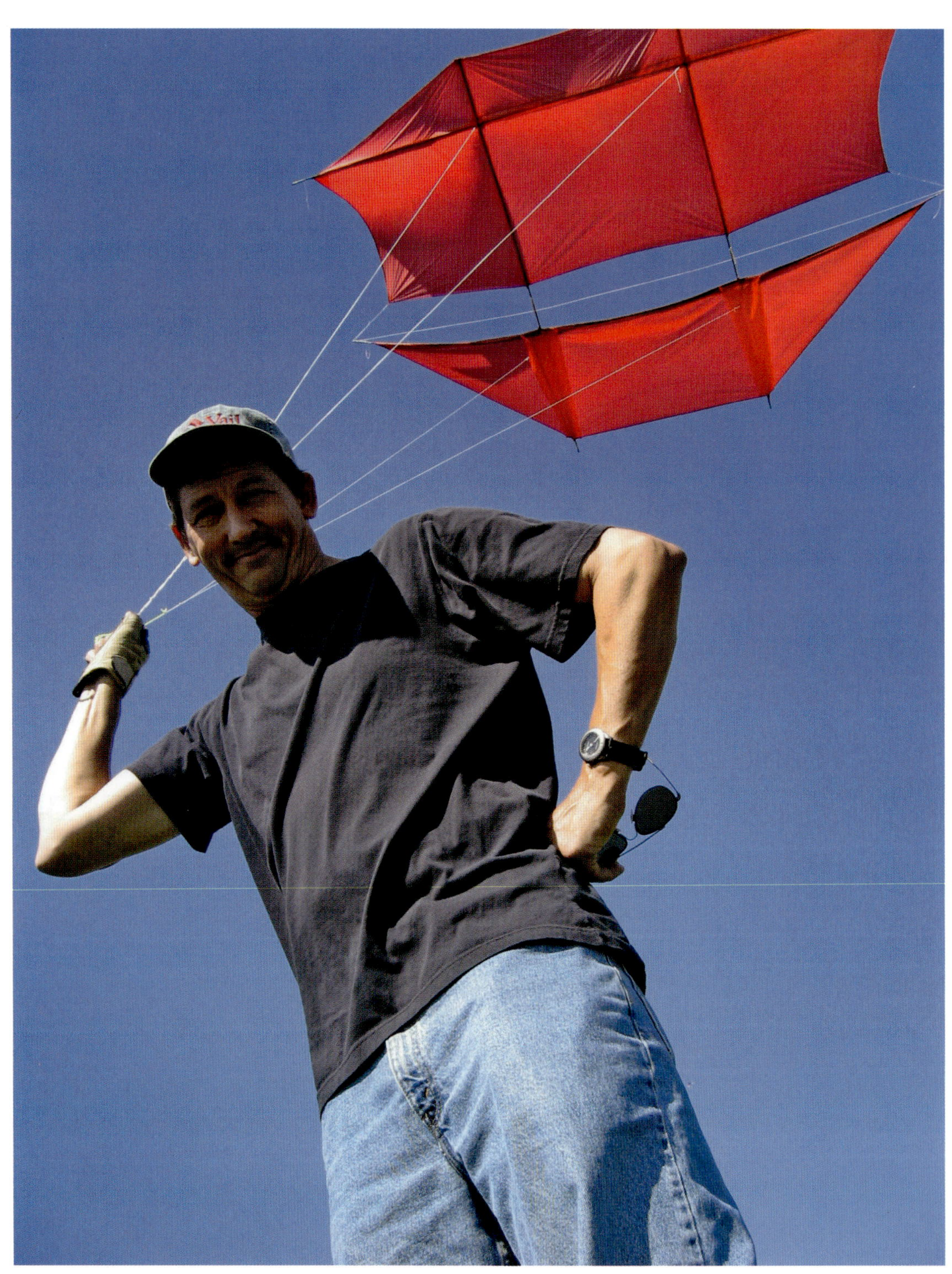

ABOUT THE AUTHOR

Cris Benton, professor emeritus of architecture and former department chair, retired from UC Berkeley in 2012. While at Cal his principal academic interests fell in the Building Science area, where he was engaged in the Building Science Laboratory for teaching and research as well as a program of post-occupancy building case studies.

Benton also harbors considerable passion for Kite Aerial Photography (KAP) and its associated historical, applied, and artistic dimensions. He began his South Bay work during a sabbatical year spent as Artist-in-Residence at the Exploratorium in San Francisco (2003), where he worked on several KAP-related projects. The work has continued under special use permits from the Don Edwards San Francisco Bay National Wildlife Refuge and the California Department of Fish and Wildlife. The resulting images documenting the salt ponds have been shown in exhibits at the Cooper-Hewitt Museum, the Exploratorium, and the Coyote Point Museum, as well as conferences and art galleries. Benton's aerial images have been used by over one hundred nonprofit agencies.

Benton takes his photographs using relatively simple handmade equipment. The camera is positioned by walking the kite around, and aimed using a homebuilt, radio-controlled cradle. Composition is accomplished in absentia as he imagines what the camera above would see. Says Benton, "Kite aerial photography appeals to that part of me, perhaps of all of us, that would slip earthly bonds and see the world from new heights. An aerial view offers a fresh perspective of familiar landscapes and in doing so challenges our spatial sensibilities, our grasp of relationships."

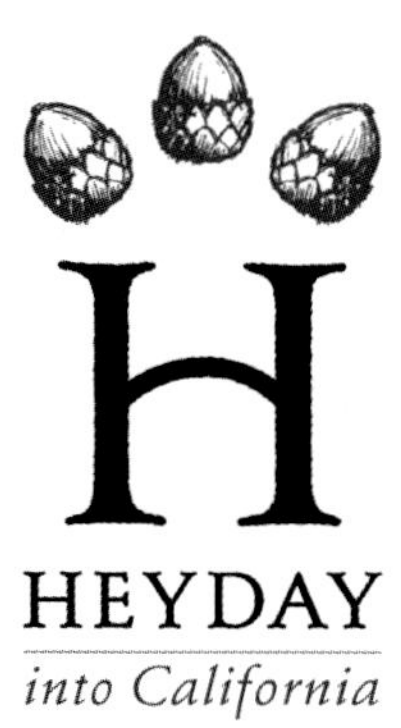

HEYDAY

into California

About Heyday

Heyday is an independent, nonprofit publisher and unique cultural institution. We promote widespread awareness and celebration of California's many cultures, landscapes, and boundary-breaking ideas. Through our well-crafted books, public events, and innovative outreach programs we are building a vibrant community of readers, writers, and thinkers.

Thank You

It takes the collective effort of many to create a thriving literary culture. We are thankful to all the thoughtful people we have the privilege to engage with. Cheers to our writers, artists, editors, storytellers, designers, printers, bookstores, critics, cultural organizations, readers, and book lovers everywhere!

We are especially grateful for the generous funding we've received for our publications and programs during the past year from foundations and hundreds of individual donors. Major supporters include:

Anonymous (3); Alliance for California Traditional Arts; Arkay Foundation; Judy Avery; James J. Baechle; Paul Bancroft III; BayTree Fund; S. D. Bechtel, Jr. Foundation; Barbara Jean and Fred Berensmeier; Berkeley Civic Arts Program and Civic Arts Commission; Joan Berman; John Briscoe; Lewis and Sheana Butler; California Civil Liberties Public Education Program; Cal Humanities; California Indian Heritage Center Foundation; California State Parks Foundation; Keith Campbell Foundation; Candelaria Fund; John and Nancy Cassidy Family Foundation, through Silicon Valley Community Foundation; Charles Edwin Chase; Graham Chisholm; The Christensen Fund; Jon Christensen; Community Futures Collective; Compton Foundation; Creative Work Fund; Lawrence Crooks; Nik Dehejia; Frances Dinkelspiel and Gary Wayne; The Durfee Foundation; Earth Island Institute; The Fred Gellert Family Foundation; Fulfillco; The Wallace Alexander Gerbode Foundation; Nicola W. Gordon; Wanda Lee Graves and Stephen Duscha; David Guy; The Walter and Elise Haas Fund; Coke and James Hallowell; Stephen Hearst; Cindy Heitzman; Historic Resources Group; Sandra and Charles Hobson; Donna Ewald Huggins; Humboldt Area

Foundation; JiJi Foundation; The James Irvine Foundation; Claudia Jurmain; Marty and Pamela Krasney; Guy Lampard and Suzanne Badenhoop; Christine Leefeldt, in celebration of Ernest Callenbach and Malcolm Margolin's friendship; Bernard and Josie Le Roy; Thomas Lockard; Thomas J. Long Foundation; Judith and Brad Lowry-Croul; Michael McCone; Nion McEvoy and Leslie Berriman; Giles W. and Elise G. Mead Foundation; Michael Mitrani; Moore Family Foundation; Richard Nagler; National Endowment for the Arts; National Wildlife Federation; Native Cultures Fund; The Nature Conservancy; Nightingale Family Foundation; Northern California Water Association; The David and Lucile Packard Foundation; Panta Rhea Foundation; Alan Rosenus; The San Francisco Foundation; Greg Sarris; William Somerville; Martha Stanley; Roselyne Chroman Swig; Swinerton Family Fund; Sedge Thomson and Sylvia Brownrigg; TomKat Charitable Trust; The Roger J. and Madeleine Traynor Foundation; Lisa Van Cleef and Mark Gunson; Patricia Wakida; John Wiley & Sons, Inc.; Peter Booth Wiley and Valerie Barth; Dean Witter Foundation; The Work-in-Progress Fund of Tides Foundation; and Yocha Dehe Community Fund.

Board of Directors

Getting Involved

To learn more about our publications, events, membership club, and other ways you can participate, please visit www.heydaybooks.com.